BUSINESS STORE DESIGN

商业店铺设计［购物篇］

深圳市海阅通文化传播有限公司 主编

中国建筑工业出版社

前言

preface

Shopping space commonly provides goods we need and works as a displaying space of goods, but is also a symbolic of social statue and identity. People always searches for the atmosphere suit for him and create his own aura while looking for goods he needed. Just like the backstage, they make themselves up to match the role they are going to play, then singing their own songs and dancing their own dances, acting their own stories.

Designs for shopping spaces is a passage connecting the backstage and forestage, people can get suitable items for them to dress up like the one they wish to be, and satisfy their daily requirements.

In this book, there are plenty of featured shopping stores from different areas in the world, including those stores selling clothing, furnishing, jewelry, and some others. They found their own position in this book and showed their functions among the computational society. This would be a visual feast and material temptation to the ones read this book.

Senior designer & Commentator Wang Yingchao

购物空间，除了为人们提供所需要的商品之外，还是商品的展示空间，也是人们社会身份与地位的象征。在这里，人们期待着属于自己的商品，同时也在空间中找寻适合自己的那种氛围，也在营造着自己的气场。犹如一个后台，人们在这里点缀自己，扮演着自己的角色，然后在人生的舞台上吟唱着自己的歌曲，演绎着自己的故事。

购物空间的店面设计，就像是后台到舞台的通道，人们从此进入找寻着适合自己的道具，从这里出去成为自己希望的那个角色，满足自己的追求。

本书介绍了来自世界各地的特色商铺，囊括了服装、家居、珠宝等各式各类购物空间，也是让它找到自己的角色，把它在消费社会中的意义展示出来，向读者呈现出视觉的享受和物质的诱惑。

资深设计师兼评论员 王颖超

目录
contents

[购物篇]

服装 鞋帽 Clothing shoes and hats

家居 Household

美发 美甲 Salon manicure

珠宝首饰 Jewelry

其他 The others

My Panda Ikebukuro PARCO

My Panda 池袋服装店

Design agency: TORAFU ARCHITECTS
Location: Ikebukuro, Tokyo, Japan
Area: 36.58 m^2
Photography: Takumi Ota

设计单位：TORAFU 建筑设计事务所
项目地点：日本东京池袋
项目面积：36.58 m^2
摄　　影：太田拓实

We performed the interior design for the Ikebukuro Parco store of the new fashion brand "my panda" centered on a "two-tones" theme. Being the second store to open after the first Shibuya Parco store, we were tasked with improving brand awareness while providing furniture and fixtures for a limited-time only shop that could be put to contribution in future iterations.

Since the location is next to the 1st floor entrance to Parco's main building which is directly linked to the East exit of the JR Ikebukuro station, we thought of appealing to commuters by two-tone painting the walls blue and white to tighten the space and by placing the giant panda logo around a pillar as in the Shibuya store. A semicircular shape was cut out from the legs of the birch plywood furniture and fixtures, such as hanger racks, tables and mirrors, and only the lower parts were painted in black to give it the cuteness factor associated with animals by making it look like a panda. The dimensions of the knock-down furniture and fixtures were established so that they may all fit in the company car in anticipation of a temporary installation at a different location and so that they may be easily assembled and disassembled by the female staff. By designing furniture and fixtures in continuity with the brand concept, we strived to create a limited-time only shop with a consistent world view.

日本 TORAFU 建筑设计事务所为新的时尚品牌 My Panda 设计了位于池袋 PARCO 大楼的零售店。My Panda 池袋服装店是继涩谷店之后的第二家零售店，以简约双色搭配为基调，希望通过店面的家具和装修来提升其品牌意识，而 TORAFU 建筑设计事务所将这些都一一体现在了室内设计中。考虑到服装店位于 PARCO 大楼主楼一楼的入口处，靠近池袋车站的东边出口，为了吸引过往上班族的目光，设计师们将墙面刷成蓝白两色，并把大熊猫的 logo 画在柱子上。桦木胶合板的展柜、衣架还有镜子等的底部被挖出一个半圆的形状，在最底部的地方用黑色漆刷成熊猫后腿的样子，增添了几分动物的可爱和趣味。商店里的家具都是可拆卸的，且它们有固定尺寸，以便于装车和临时装卸，即使是店里的女员工，装卸起来也并不费力。

Sports Authority – Mark Is Minato Mirai

"Mark Is 港湾未来" Sports Authority 概念店

Design agency: TORAFU ARCHITECTS
Location: Minatomirai Kanagawa, Japan
Area: 378.3 m^2
Photography: Daici Ano

设计单位：TORAFU 建筑设计事务所
项目地点：日本神奈川
项目面积：378.3 m^2
摄　　影：阿野太一

We established a concept shop within the Sports Authority located in the Mark Is Minato Mirai mega shopping mall.

Built around the theme of running and basketball, approximately half the store is dedicated to Nike products while the other half offers a selection from a number of other brands. We were tasked with creating a select shop carrying various brands while providing a clear spatial separation between Nike products and other brands, and incorporating a sporty atmosphere. The store is divided in two areas by a large gate, with Nike products at the back and other miscellaneous brands near the entrance. The spatial structure of the store was converted in order to shift the scenery. The gate progressively frames and enhances each area and the store as a whole amid a lively environment. The flooring, walls and fixtures are made from elements found in the sports field. We incorporated the essence of sports into the shop with such elements as field-and-track rubber chip flooring material and track lines, cemented excelsior boards, such as those found in sports arenas, and furnishings like a bench press, etc. While sharing the same materials, a different color palette was used in the Nike area to distinguish it from the other brands. Even if the two areas are clearly demarcated, the overlapping fixtures and track lines give both areas a sense of unity as a single shop.

Sports Authority 概念店位于日本横滨"Mark Is 港湾未来"大型商业区。

围绕跑步和篮球的运动主题，店里一半的商品都来自 Nike，除此之外，其他品牌的服装和跑鞋也有不少。按照业主的要求，设计师将室内空间分成两个区，后面的区域专门摆放 Nike 的产品，靠近入口的区域则用来放置其他品牌的产品。两区中间用一道大门隔开，既有区分又相互连通，并通过室内设计，将概念店打造成一个具有运动气质的空间。设计师还将运动场上的一些元素运用到店内的地板、墙面和家具中，如画成跑道的地板，形象的跑道线和水泥地面足以以假乱真；还有运动用的卧推凳，真实得让人恍如正在上一节体育课。店内的两个区域都用了相同的材料，但为了区分，两区的色调各有不同。尽管如此，同样的家具陈设和相连的跑道线，又将两个分区合二为一。

NIKE

Baby Walz Exhibition

贝沃兹母婴品牌展示厅

Design agency: Shanghai Window Design
Designer: Cheryl Lee
Location: Wenzhou, Zhejiang
Area: 560 m^2

设计单位：上海尚窗室内设计有限公司
设 计 师：李秀儿
项目地点：浙江省温州市
项目面积：560 m^2

Babywalz is a brand of maternal and infant supplies with rich categories. In order to know the culture of this brand better and to be inspired, designer looked into the website and product catalogs, and find out elements like birds, clouds, hills and mushrooms are frequently used. It occur to designer to draw a fantasying fairy-tale picture with the brand color—Sapphire, which was finally adopted.

The surface of the ground floor is small, so it was designed into comprehensive exhibition area. There are ropes between trees to hang baby clothes, small mushrooms and colorful stones made by clothing are scattered around the paths while lovely clouds are flowing on the sky, all these elements made up an ease and comfortable atmosphere and also interpreted the products. The first are on upper floor is designed into a nest, solving the predicament caused by narrow entrance and huge walls by building a small house with fantastic and fairy look.

The overall planning adopts color as a tool to zone each area, being independent but could blend. Sapphire and white are the main colors, while each zone has a specific color with the same lightness and purity, creating a naiveté space. At last, each two zones are linked by intermediate colors.

贝沃兹是一个产品类别丰富的母婴品牌。设计师从官网和产品目录手册中了解品牌文化和汲取灵感，惊喜地找到小鸟、云、山丘和蘑菇元素，以及品牌色－粉蓝，马上联想到一幅幅梦幻的童趣画面，灵感浮现，很快就确定把梦幻的情景化设计作为整体概念。

原建筑一楼面积小，被设计成综合展示区。一棵树与树间的绳子上晾晒着宝宝衣服，路边上有很多布艺小蘑菇和彩色的小石头，天上飘着可爱的云朵，惬意而亲切的氛围，引人入胜，又将主要产品线都展示出来。二楼展厅第一区域，设计师将其设计成小鸟的家，小房子的结构巧妙化解狭窄入口和超大结构墙堵在入口的窘境，带来童话般的梦幻感。

整体规划上，设计师用色彩作区分，每个区域之间既能独立又能融合。首先选择粉蓝色和云朵的白色作为贯穿整体的基色，再把每个大区域赋予一个主色，每个颜色都是统一明度与纯度，都有天真烂漫的感觉，每个区域之间又用一个关联色相互融合。

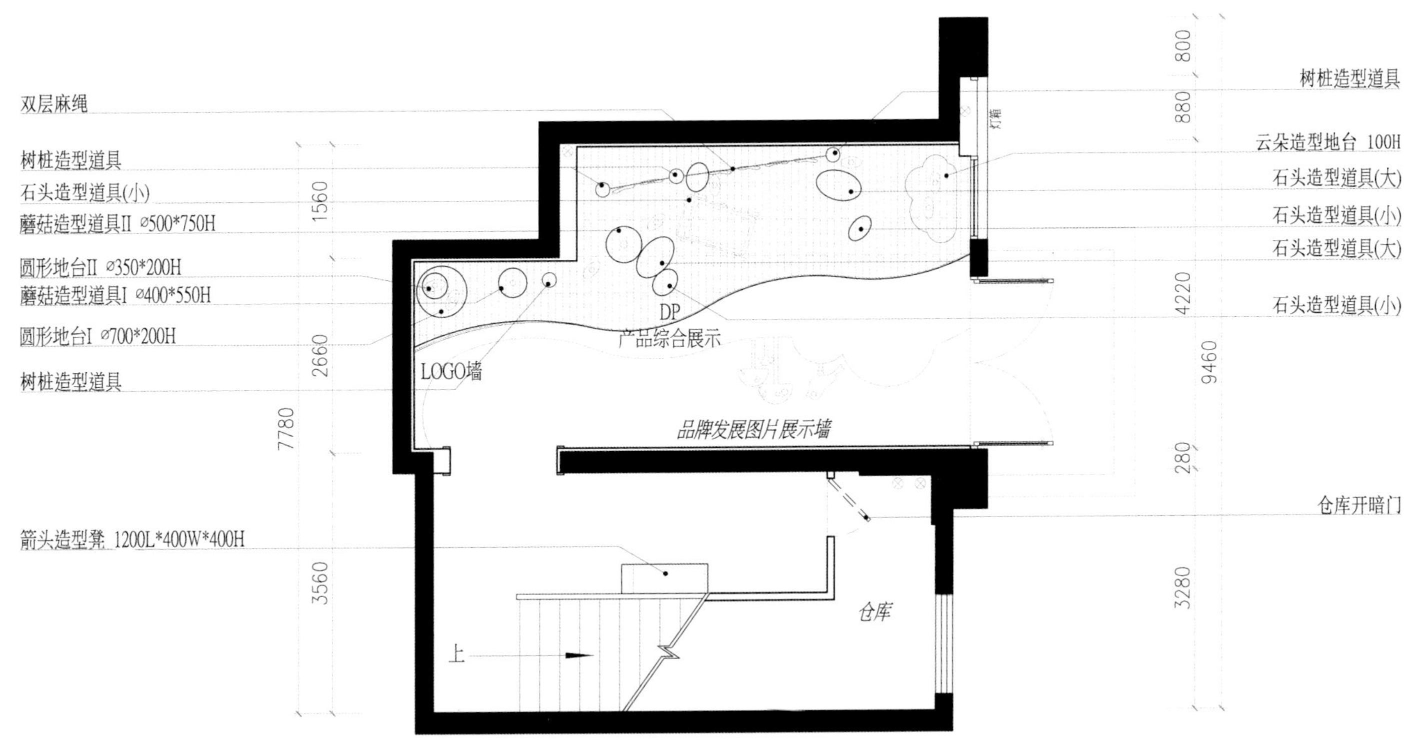

一层平面布置图

welcome to the family

GIRLS
kids
BOYS
BOYS
babywalz

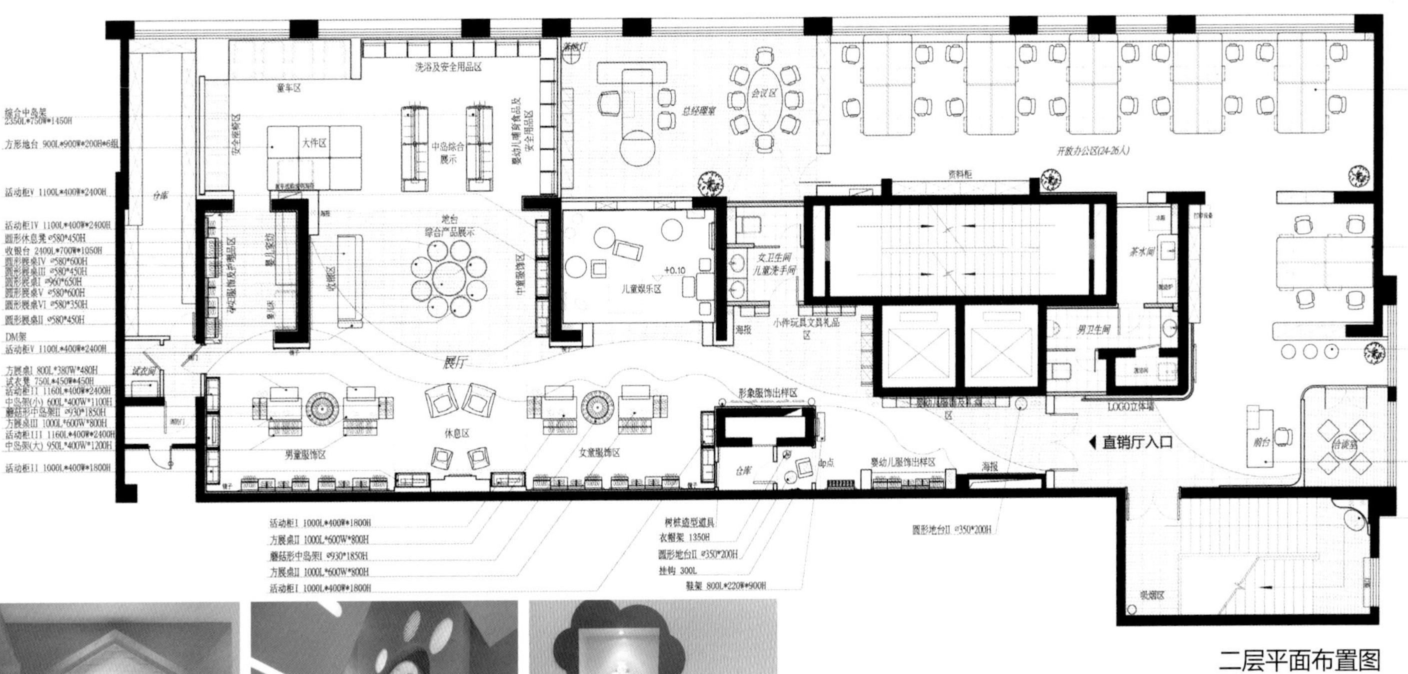

二层平面布置图

Giuseppe Men's Store

乔治白男装专卖店

服装鞋帽
Clothing shoes and hats

Design agency: Shanghai Window Design
Designer: Cheryl Lee
Location: Wenzhou, Zhejiang
Area: 268 m^2
Photography: Rocky Wang

设计单位：上海尚窗室内设计有限公司
设 计 师：李秀儿
项目地点：浙江省温州市
项目面积：268 m^2
摄　　影：Rocky Wang

This case is designed for Giuseppe Men's store, which is a historic brand. This is an European brand focus on the suits, thus European classic style fits to interpret the culture of suits and highlight the Italian characters of this brand.

Space layout is designed into propelling ways, changing the asymmetric space into symmetric space while arched door functions as a transition. There are three regions: sales area in the first region; checkout area in the second area which is also a sales area; the third area is a fitting area appears like a living hall in European buildings, being private like a VIP.

Classic European materials are used in this case, including carved cherry wood, yellow marbles and gypsum and so on. Wood seems to be simple, but the decorations and installations are great, such as cabinet made by leathers, specially designed marble patterns, textured glasses, custom made carpets with particular colors and patterns, all these make up a gorgeous space.

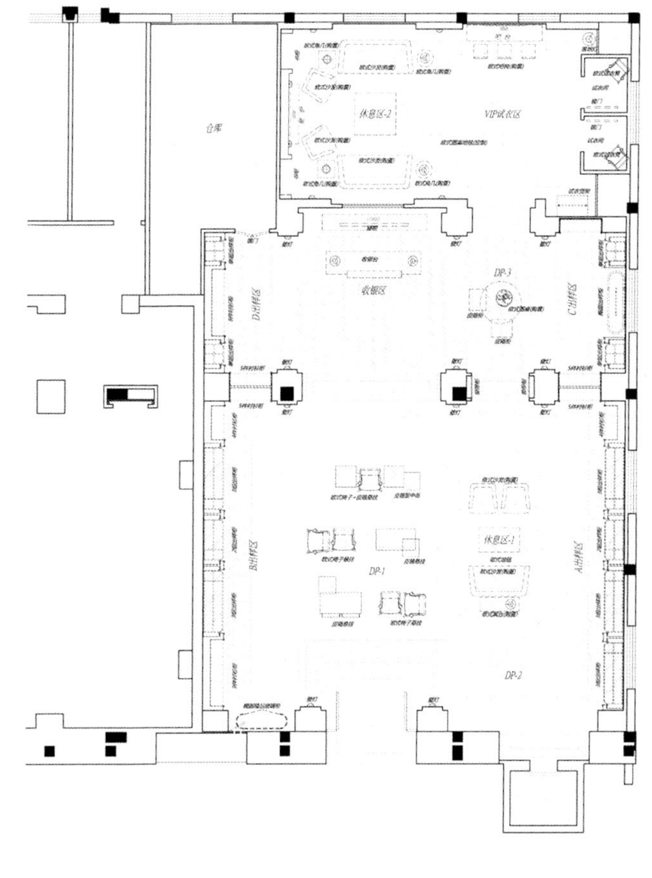

平面图

本案是乔治白男装品牌专卖店，品牌具有悠久的历史。该品牌产品以正装为主，正装西服来自欧洲，欧式经典能很好地诠释西服的文化，并突出品牌来自意大利的出身。

空间布局采用递进式设计，把并不对称的空间规划成对称空间，用拱形门作为递进空间的过渡点。第一区域最大，作为卖场区，第二区域为收银区，同时也是卖场区，第三区域为试衣区，设计成欧洲家庭客厅式，私密而体现贵宾感受。

在选材上，采用欧式经典的材料，樱桃木线条及雕花、米黄色系大理石、石膏线条等，木材虽简单，但是辅材与软装却丰富，有真皮制做的皮箱柜、特别设计的大理石地花、肌理玻璃、色彩与花型独特的定制地毯等。

Vigoss Design Studio

VIGOSS 服装店

服装鞋帽
Clothing shoes and hats

Design agency: Zemberek Design Office	设计单位：Zemberek 设计事务所
Designer : Başak Emrence, Şafak Emrence, Başak Bakkaloğlu	设 计 师：赛法克•艾伦斯、巴塞克•艾伦斯、巴塞克•巴克卡尔奥卢
Location: Istanbul, Turkey	项目地点：土耳其、伊斯坦布尔
Area: 500 m^2	项目面积：500 m^2
Photography: Şafak Emrence	摄 影：赛法克•艾伦斯

The demand was creating a perception of "non-belonging" in the space, to highlight the difference between existing function and the new one, also maintaining the balance, between using the attractive and colorful interior workspace as a showcase and preserving the privacy of the designers. Another demand that shapes the concept was a multi-functional area which serves as a design office and a showroom.

Architectural team aimed to meet all demands with a single shell. The shell is: 1, creating a sense of "dissimilarity" by its form which is contrary to the form of existing spaces; 2, developing a perception of "addition" to the former function of the area; 3, providing the connection with existing spaces and its users by a permeable layer; 4, serving both as a seperation and display, continuing around the office area.

By the permeable feature of shell, division between inner and outer spaces becomes indistinct, in addition, continuity of connection with existing spaces and its users is ensured. Form of the cnc slabs, offers variable visibility levels of inside, to users walking around the shell, thus brings controlled privacy to the users inside. With shadows generated by lighting pass through the slabs, the shell also creates a visual value for the surroundings.

In accordance to the concept of the shell, working space is designed as a monolith form , on the purpose of creating a perception in terms of "integrity". Designer's desk is formed as a massive concrete plate in the middle, supported by columns.

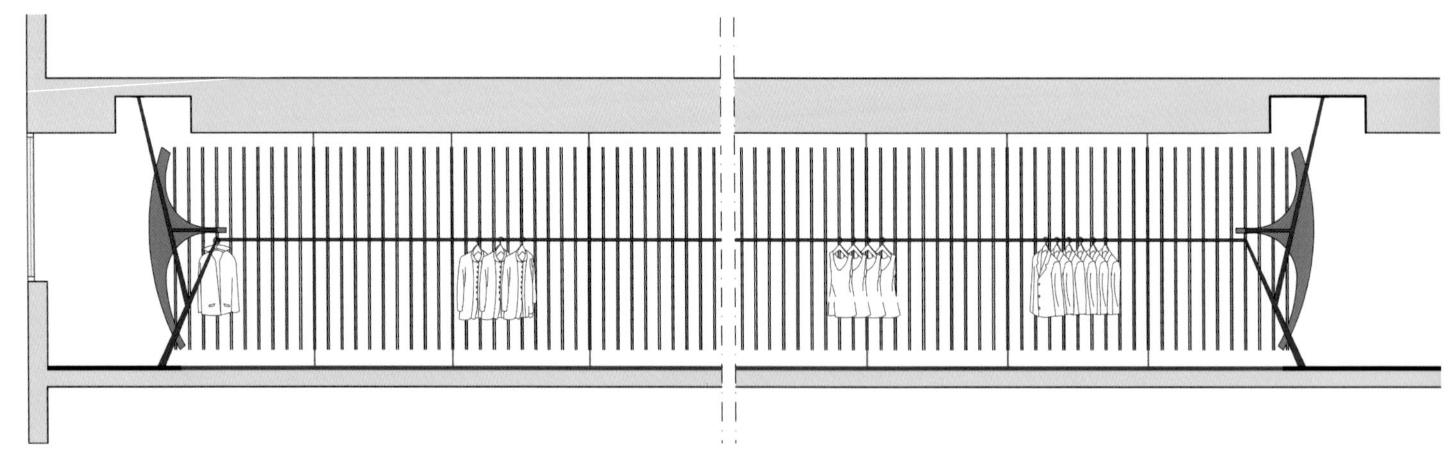

平面图

客户要求在空间内设计一种“非从属”空间的感觉，既要强调新旧空间之间差异，同时也要保证两者的平衡性。在设计中，将色彩鲜艳的富有吸引力的工作区域设为展示部分，同时保证了服装设计师的个人隐私性。甲方的另一个要求也是设计成型的另一个因素即空间的多功能性：既是一处设计的办公室，同时也是一个展示的空间。

设计团队意图利用一种“壳”的手法来解决所有的设计要求。这个“壳”有很多好处：1. 通过其自身的形态与当下已有空间形态形成对比和差异；2. 给之前的空间营造一种新空间的感觉；3. 给空间的使用者和空间本身提供一种可渗透的关联；4. 在办公区域的附近营造一处既是分离又可供展示的区域。利用“壳”装置的渗透特性，内外空间的区别不是特别明确，此外，与已经存在的空间及其使用者的联系也得到了保证。而这些数控切割的板条，给内部的使用者提供了多种多样的可视度变化。就算在装置附近有人走来走去，也能给内部的使用者一定的私密性。

为了营造一种整体的感知力，与“壳”装置的造型相协调的是设计成整体的工作空间。设计师的工作台的形式都是巨大的水泥平台，并坐落在空间的正中，由柱子承载受力。

Pilar Vidal Shop

皮拉尔·维达服装店

Design agency: Vicente Vidal Estudio　设计单位：维森特•维达工作室
Designer: Vicente Vidal　设 计 师：维森特•维达
Location: Tenerife, Spain　项目地点：西班牙特内里费
Area: 33 m^2　项目面积：33 m^2
Photography: Sepia Estudio　摄　　影：Sepia 工作室

The prestigious fashionable designer Pilar Vidal, art director of the company, has trusted us this ambitious project that tries to transmit the peculiar vision of both signatures "Pilar Vidal" and "Modenina", both of the above mentioned designer. Pilar Vidal is inspired by the light, the color and the breeze of the Mediterranean Sea. She travels across the time to offer us a different vision of the femenine that does not leave indifferently.

She wants to invite and attract clients to get inside this space and get immersed in a great dreamed dressing room, as if they were in house, a dressing room that transmit a world of dreams, fantasy and happiness where the amusement and the pleasure are possible... Fuchsia, fitting rooms with great mirrors remind the artist dressing rooms. A very delicate lighting with decorative elements gives a soft, evolving and "retro" atmosphere.

PILAR VIDA

著名设计师皮拉尔·维达，亦是该公司的艺术总监，对于这一项目给予了极大的信任及支持，尤其是在设计公司试图在皮拉尔和莫德妮娜的前面嵌入设计中的时候。皮拉尔的设计灵感来自于光、色彩以及地中海吹来的徐徐海风。她经常过来拜访，并且给设计师留下了热心温暖的印象。

她想要吸引并邀请客户进入空间，沉浸在这一个梦幻的空间，如同一间容纳了时间、所有的美梦、幻想和幸福的房子，在这里，可以找到所有的快乐和欢愉。带有试衣大镜的紫红色试衣间展示出独特的艺术氛围。装饰灯泡里面散发出的柔和微光为空间增添了一丝“复古”的韵味。

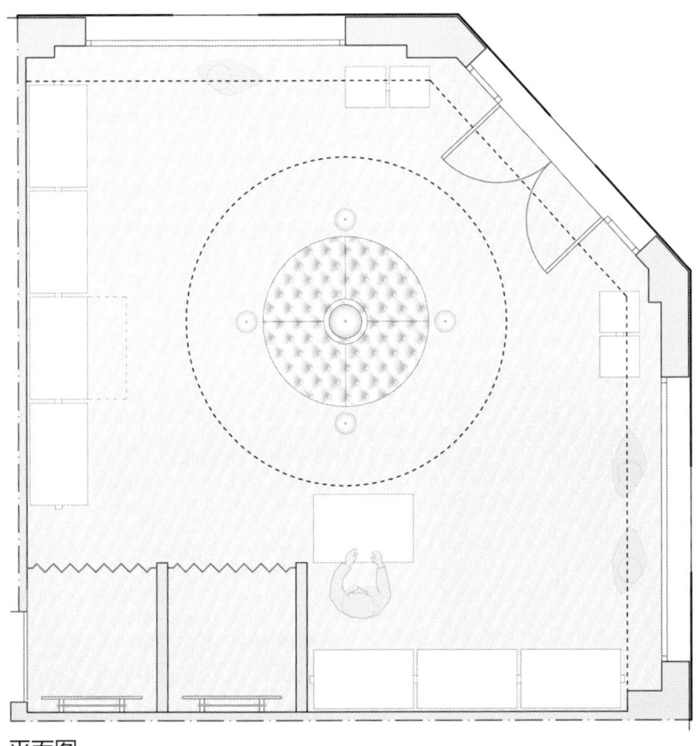

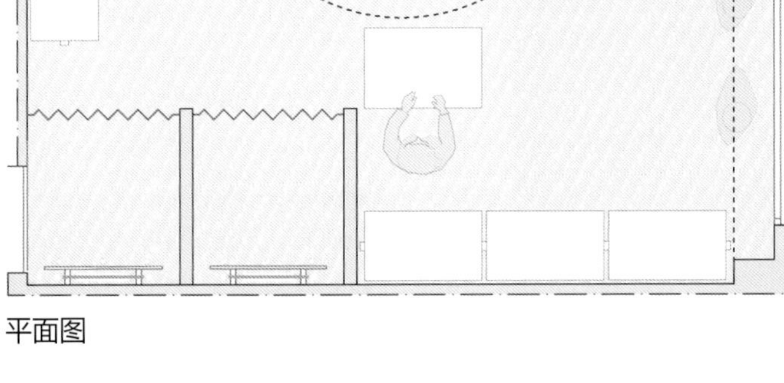

平面图

Vince Camuto Grand Central Terminal

文斯·卡莫多中央火车站店

Design agency: Sergio Mannino Studio
Designer: Sergio Mannino, Francesca Scalettaris, Alessandra Prezzi
Location: Manhattan, USA

设计单位：塞尔吉奥 • 曼尼诺设计工作室
设 计 师：塞尔吉奥 • 曼尼诺、弗朗西斯卡 • 斯卡莱塔里斯、桑德拉 • 佩里兹
项目地点：美国曼哈顿

The Vince Camuto flagship store in Grand Central Terminal New York is the first retail location dedicated to the US brand. It is designed to sell footwear, handbags, outerwear and accessories. This sophisticated space balances modern product display elements with a sharp material palette to frame, reflect, and layer Camuto's edgy feminine style.

The studio faced structural challenges when designing the store: the floor level of the store was higher than the rest of the terminal and there were also several columns that blocked the view, making it difficult to place fixtures around the store. The architect also used materials like white lacquer, polished chrome metal, and a ceramic floor called Esko by Impronta Italgraniti. The color palette takes on different shades of gray for a balance of modernism and sophistication.

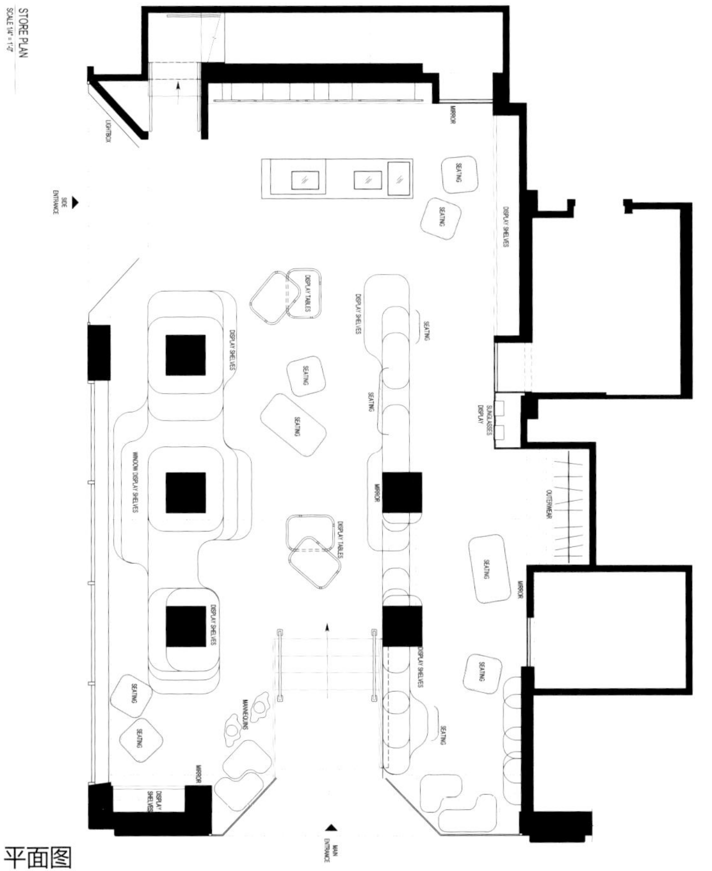

平面图

位于中央火车站的文斯·卡莫多旗舰店是塞尔吉奥·曼尼诺工作室对美国本土品牌设计的第一次尝试。该品牌涉及鞋类、手提包、外套以及配饰等各方面。展示在锐利材料板铸就的框架上的时尚产品展现出前卫女性风格，在这一复杂的空间里面达到了平衡。

工作室在设计这个商店的时候面临结构性的挑战：商店的地板水平高于车站的其余部分，同时还有一些柱子挡住视线，加剧了设置商店设备的难度。还采用了白漆材料、抛光铬金属和艳皕达生产的一种陶瓷地板。彩色木板上深浅不一的灰色平衡了空间的现代性和复杂性。

The Flagship Store of Brioni in Frankfurt

Brioni 法兰克福旗舰店

Design agency: Park Associati
Location: Frankfurt, German
Area: 320 m^2
Photography: Andrea Martiradonna

设计单位：Park Associati
项目地点：德国法兰克福
项目面积：320 m^2
摄　　影：安德里亚•马蒂拉多那

The men's high fashion Italian brand, of group Kering, has entrusted to Park Associati the implementation of the new concept for the interior design of the international boutiques of which the store in Frankfurt is currently the most extensive example of square footage in the world (320 sqm of two floors). The distribution of the various product areas expressed in this space the classical division, designed by Park Associati to maximize the brand's products, and correspond to the expectations of the customer to welcome the customer in an elegant and linear space, in which the materials enhance their classic taste of a brand that was born in the city of Rome, the heart of European classicism.

The space on the ground floor, double height, houses the rooms dedicated to accessories and leather accessories as well as the leisure. The large piece of furniture for the display of accessories occupies the wall that separates the entrance from the staircase leading to the upper floor, growing to full height. Particular importance has been given to the design of the stair: real architectural element that characterizes this store. Breaking the geometric linearity of the space, the stair grows developing itself around acute angle between the ramps.

Upstairs, the rooms are dedicated to footwear, formalwear and taylormade and to the VIP Room area. In the VIP Room, the concept of elegance and exclusivity that characterizes the brand is emphasized by the presence of a lounge area fitted out with unique furniture aimed to the Su Misura. A large closet opens up to show all finishes and fabrics used for customizing clothes. The wooden ceiling is decorated with a series of glass lamps on ribbed design. The atmosphere is made even more exclusive by finishing that, even in continuity with the shop area, emphasizing the craftsmanship of the work. It ' was deliberately chosen to use fewer materials, with careful attention to detail in order to communicate a discreet luxury and precious.

Park Associati 受邀为意大利高端时尚品牌 Brioni 法兰克福旗舰店提供设计，将这个 320m² 的广阔空间打造成高端国际精品店。店内的商品分区设置透着经典，既充分利用了空间，又为客户提供了一个线条利落又不失优雅的购衣环境。室内设计还采用不同的材料以增加品牌的古典品味。首层部分空间层高为双层高，主要安排了皮革配饰区和休闲区。到顶的货品架犹如一道装满珍品的高墙，将入口处和楼梯分隔开来。楼梯引领着客户上到第二层，设计师别出心裁地运用了一些建筑元素，使整个楼梯围绕一个锐角形成折转，打破了整个空间单一的线条感。二楼主要有鞋区、正装区和贵宾室。贵宾室的设计注重优雅和专享的概念。里面的大衣柜展示了用于定制服装的各种衣料，配合独特的家具，透露出一种“量身定做”的尊贵气息。木质的顶棚还用带棱纹的玻璃灯装饰。

法兰克福旗舰店风格优雅，精挑细选的材料和精心的设计将其注重细节和工艺品质、低调奢华的特点一一展现。

Glassons

Glassons 女装专卖店

Design agency: Landini Associates	设计单位：Landini Associates
Location: Sydney, Australia	项目地点：澳大利亚悉尼
Area: 223 m^2	项目面积：223 m^2
Photography: Sharrin Rees	摄　　影：谢林 • 里斯

Landini Associates collaborated with New Zealand fashion brand Glassons to re-launch the retailer on Australian shores with a new retail format at Macquarie Park and Bondi Westfield. The concept has since been rolled out to Queensland and various locations across New Zealand.

The brief was to better present Glassons' great value product such that collections were more intelligible and the perceived quality higher without reducing SKU's. In response, Landini chose to create more walls without merchandising the real ones, and hid the changing and stock rooms behind a mirror on one wall and projected film floor to ceiling on another.

Landini Associates has challenged traditional display methods by removing all the product from the walls and creating wardrobe-like display units throughout the store. This design approach eliminates the cluttered feel of many high-street fashion stores, creating a clean look whilst cleverly maintaining a high volume of product. In fact the design allows for 50% more products on the floor and this location is now Glassons' best trading location per square meter.

The designers wanted to create a clean, sharp and fresh interior with a touch of urban feel. Featuring a warm white palette, the design utilizes light oak timber, concrete, mirrors and a white floor.

Also unique to the new retail format is the celebration of the local Glassons identity. The floor to ceiling projection was designed to showcase local imagery and create a sense of locality to the brand in each location. This brings authenticity into the Glassons retail experience and is reinforced by brand identity words used throughout the store: "Made for here."

Success was immediate with stores sales consistently exceeding plan.

Landini Associates 为新西兰时尚品牌 Glassons 在澳洲设立的零售店提供了独特的设计。整个设计的重点，在于在不减少库存单位的情况下，能够更好地展现 Glassons 产品的优势，使其拥有更高的品质认知度。因而设计师有意将室内打造成一个清新且干净利落，又不失城市感的空间。整个空间的主色调是温暖的白色，还采用了橡木、混凝土、镜子和白色地板。设计师向传统的服装店展示方式发起挑战，未使用服装挂在墙上的方式，而是采用了衣柜式服装展示架，既消除了商业街时装店的凌乱感，又保证了一定的展示空间。店面的落地窗玻璃，使新上的服装得以处在最显眼的位置。到顶的大屏幕占满了一面墙，用于展示品牌的地方形象，增添各地的地方感，使顾客的购物体验显得更加真实。正如店内随处可见的品牌标语所要表达的一样：美衣华服，此处尽享。

GLASSONS
SHORTS
$24.99

MADE
OF
HERE
MADE
OF
HERE

GLASSONS

Scrapbook - Jeanasis

时尚品牌店

Design agency: Sinato Inc.
Location: Tokyo, Japan
Area: 295.6 m^2
Photography: Toshiyuki YANO

设计单位：Sinato 建筑设计有限公司
项目地点：日本东京
项目面积：295.6 m^2
摄　　影：矢野纪行

The project is a street-level shop for fashion brand SCRAPBOOK (JEANASIS).
In order to use both levels of the previously separated first and second floors, a portion of the floor slab was removed and a new stair created. Considering the position of the existing beams and the required size of the stairs, we were limited to positioning the staircase in the center of the shop floor. As such, we considered the stair as not just a simple device for going up and down, but as having a symbolic existence reacting to a variety of activities, creating the characteristic of vertical circulation within an accumulation of the functions of product shelf, display, fitting room and bench.
Additionally, in the entrance area, we removed the existing door assembly and created a new 4 meter width opening featuring a single giant sliding door. During business hours, the large door is always open, and when stopped in its open position, the door creates a new "framed" display of the adjacent space on the inside of the shop.

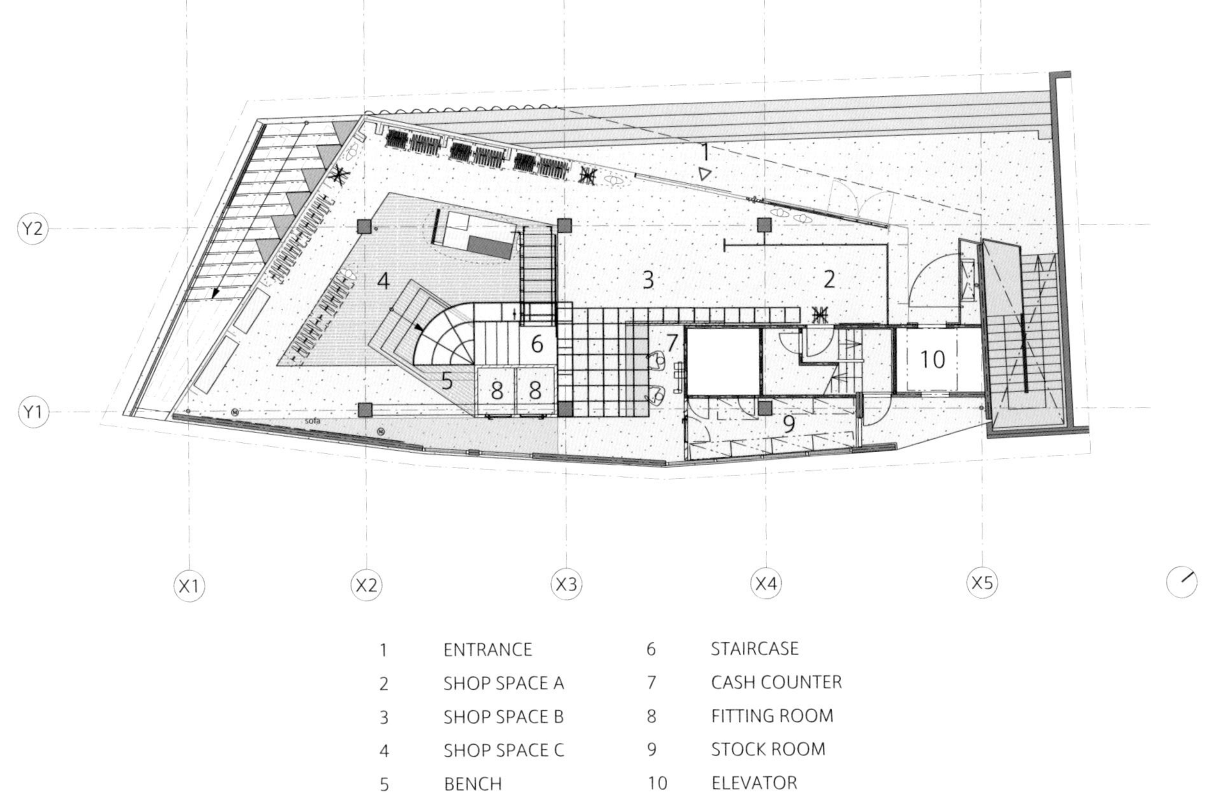

平面图

该项目是为时尚品牌 Scrapbook– Jeanasis 所设计的临街门店。

为了同时使用被隔断的一二两层，楼板的部分隔断被拆除，取而代之的是一副新楼梯。针对现有房梁的位置以及所需楼梯的尺寸，楼梯间被限制在了卖场的中央位置。因此，设计师让楼梯的角色不限于简单的上上下下的装置，更是一种象征性的存在，对各类活动作出相应的反应。同时也集货架、试衣间、展台和台架等功能于一身，从而展现出垂直环流特色。

此外，入口处的现有门组件被拆除，并创造了一扇 4 米宽的新滑动门。营业时，大门一直处于开启状态，当其停留在固定的开启位置时，将店内临近门的空间框起来又形成了一个新的风景。

(JEANASIS)

FITTING
2F BOOK
STATIONERY
COSMETIC
SOUVENIR
FOOD

Jaeger Store

耶格服饰专卖店

Design agency: UXUS	设计单位：UXUS 设计事务所
Location: London	项目地点：英国伦敦
Area: 257.2 m^2	项目面积：257.2 m^2
Photography: Mark Davison	摄　　影：Mark Davison

UXUS was commissioned by luxury fashion brand Jaeger to develop a new retail design concept at the Jaeger store on London's Kings Road. The new retail platform expresses the Jaeger ethos of modernity, quality, and understated British confidence, communicating Jaeger's heritage in new and exciting ways. Inspired by everyday moments of artistry, the Jaeger store invites customers to discover a series of inspirational yet effortless moments of elegance. Flexible stenographic table and wall presentations showcase the quality of each Jaeger piece with understated luxury.

Immaculate attention to detail ensures a tactile, responsive environment, where natural materials invite the customer to linger in comfort. With supremely high quality finishes, the resulting design incorporates features such as oak flooring, woven and deep pile carpet to delineate specific destinations within the store; black walnut wall panels and table tops along with woven textiles for wall finishes, curtains and upholstery; and iron and light bronze metal accents.

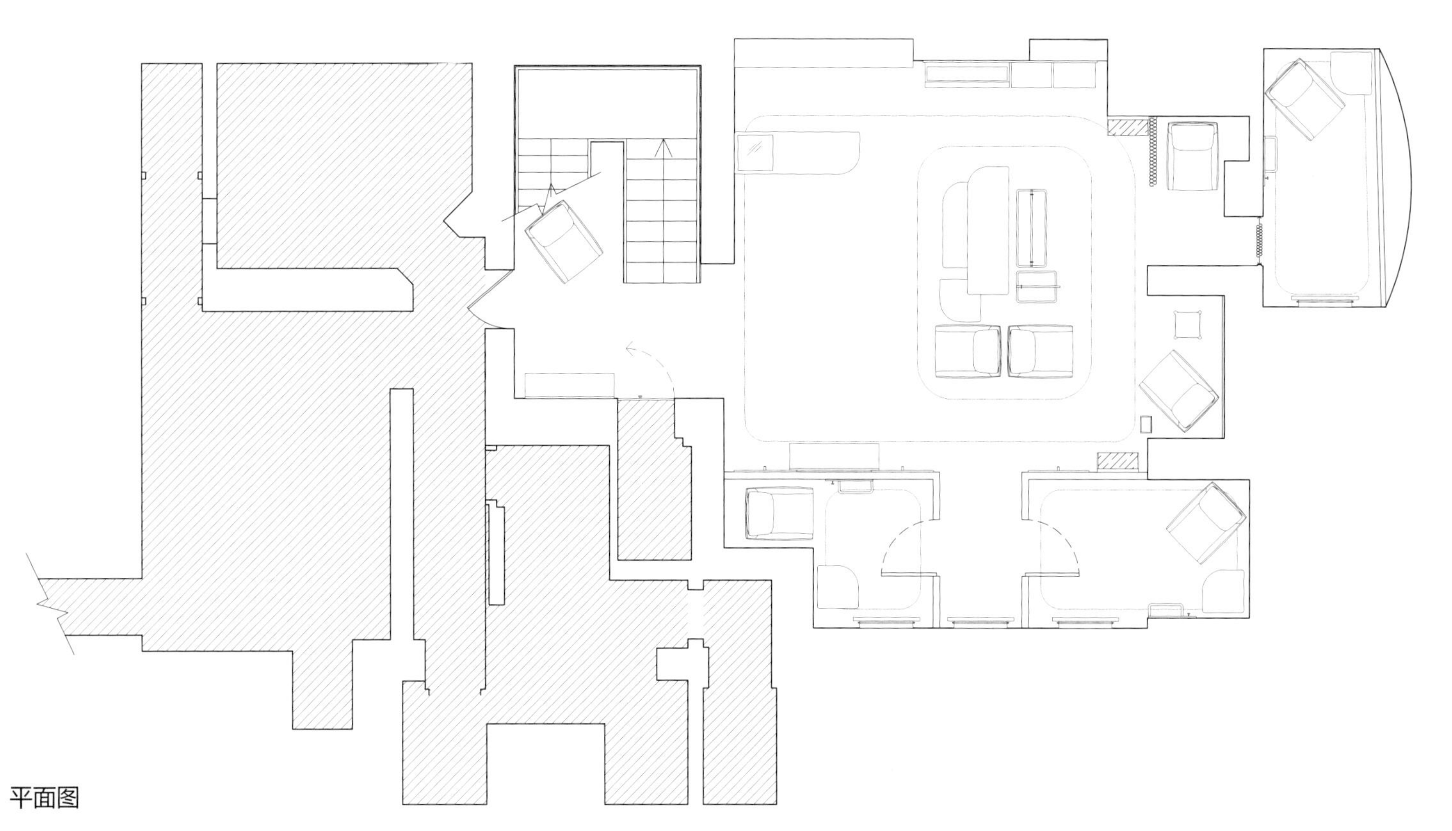

平面图

时尚奢侈品牌耶格委任 UXUS 为其在伦敦国王街的门店，开发设计新的零售店理念。这个新的零售平台表达出耶格的现代感、品质感和对英国自信的诠释，以一个全新的、振奋人心的方式表达出耶格的传承特点。受到耶格丰富档案的启发，签名店邀请顾客探索发现一系列发人深省的高雅瞬间。灵活的透视绘画布景桌子和墙壁展示品，展示出每个单品的品质和奢侈感。

对细节的极致关注确保了触觉上有回应的环境，在这个环境中，天然的材料吸引顾客舒适地徜徉其间。极高品质的面漆，最终设计结合了诸如橡木地板、编织长毛绒地毯这样的特色、黑色胡桃木墙壁面板和沿着纺织品排列的顶部桌子、作墙壁的装饰、窗帘和家具装饰用品、铁器和浅青铜金属元素。

MADE IN GREAT BRITAIN

JAEGER GENERATIONS
JAEGER

Qela Store

Qela 品牌店

Design agency: UXUS	设计单位：UXUS 设计事务所
Location: Doha, Qatar	项目地点：卡特尔多哈
Area: 400 m^2	项目面积：400 m^2
Photography: Adrian Haddad	摄　　影：Adrian Haddad

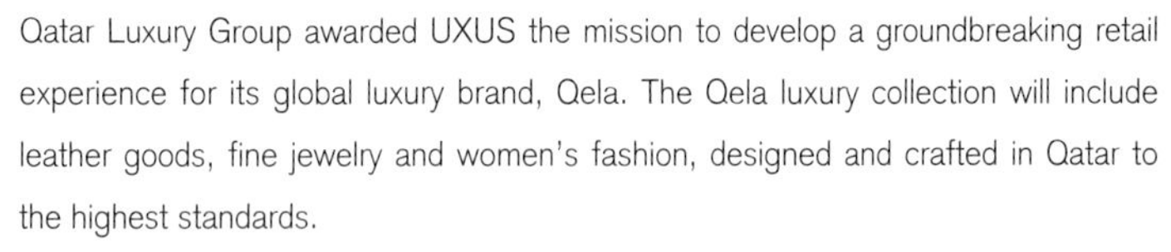

Qatar Luxury Group awarded UXUS the mission to develop a groundbreaking retail experience for its global luxury brand, Qela. The Qela luxury collection will include leather goods, fine jewelry and women's fashion, designed and crafted in Qatar to the highest standards.

The new store strikes a delicate balance between the brand's cultural heritage and its progressive spirit, immersing customers into the world of art, culture and design. Following this concept, the boutique's design blends the atmosphere of an intimate salon with the dynamism of a world-class art space: inviting clients to explore a continuously changing gallery of crafted luxury products and fine art. Aesthetically, the Qela experience has its roots in travel and embodies its native country's aesthetics of quiet, pure and natural landscapes. Undulating forms and flowing curves echo majestic Qatari vistas, punctuated with rich, luxurious materials and a palette of elegant desert tones to highlight the Qela collection.

全球奢侈品品牌卡塔尔邀请 UXUS 为其旗下的国际奢侈品牌 Qela 制定了开创性的零售。该品牌的产品包括皮具、珠宝和时尚女装，都是以卡塔尔的最高标准设计和制作的。

设计师要把品牌商店的设计在品牌文化与进步精神之间作一个微妙的平衡，使顾客进入其中，犹如进入一个艺术世界。根据这一概念，这一精品设计融合了艺术空间的活力并且有一个亲密的沙龙氛围：邀请客户探索不断变化的奢侈品。从外观上看，商店拥有本土的安静和纯洁，并且结合本土的自然景观。起伏的形态和流动的曲线，豪华的材料和色彩，让这里显得优雅丰富。

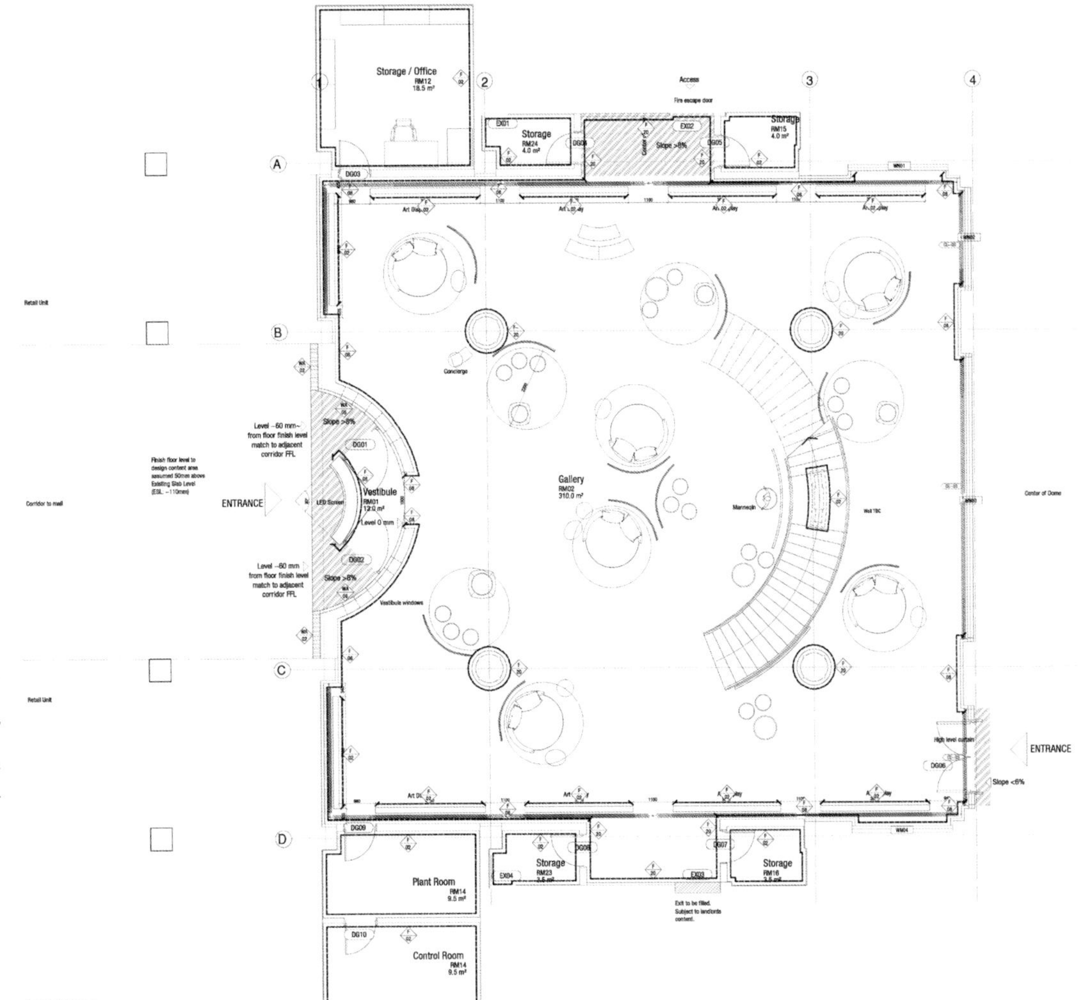

平面图

PLEATS PLEASE+BAO BAO ISSEY MIYAKE Narita Airport

三宅褶皱＋三宅一生 BAOBAO 成田机场品牌店

Design agency: MOMENT
Designer: Hisaaki Hirawata, Tomohiro Watabe
Location: Narita, JAPAN
Area: 34 m^2
Photography: Fumio Araki

设计单位：Moment 设计事务所
设 计 师：Hisaaki Hirawata、Tomohiro Watabe
项目地点：日本成田
项目面积：34 m^2
摄　　影：Fumio Araki

ISEEY MIYAKE opens new combined store introducing PLEATS PLEASE and BAO BAO inside Narita international airport. This store simply embodies each characteristic to stimulate the easy understanding for the various passersby. The corrugated boars represent the original clothes-form and lightness for PLEATS PLEASE. The lighting lines making the triangular pieces on the glass screen represent the icon of BAO BAO. Only two minimal patterns construct this store. Two different brands can be mixed without any border lines as the representation is focused on each strong point.

三宅一生在成田国际机场开辟了新的组合店，着重引进三宅褶皱及BAOBAO两个支线品牌，该店只采用了品牌各自的特色来引起各色路人的注意。褶皱波纹代表着三宅褶皱的服饰特色和轻便，而在玻璃幕墙上形成三角形的照明线则象征着BAOBAO的标志。整个专卖店都是由这两个极小的图案构成，像这样将两个完全不同的品牌无缝地糅合在一起，只有集中体现各自闪光点才是最好的选择。

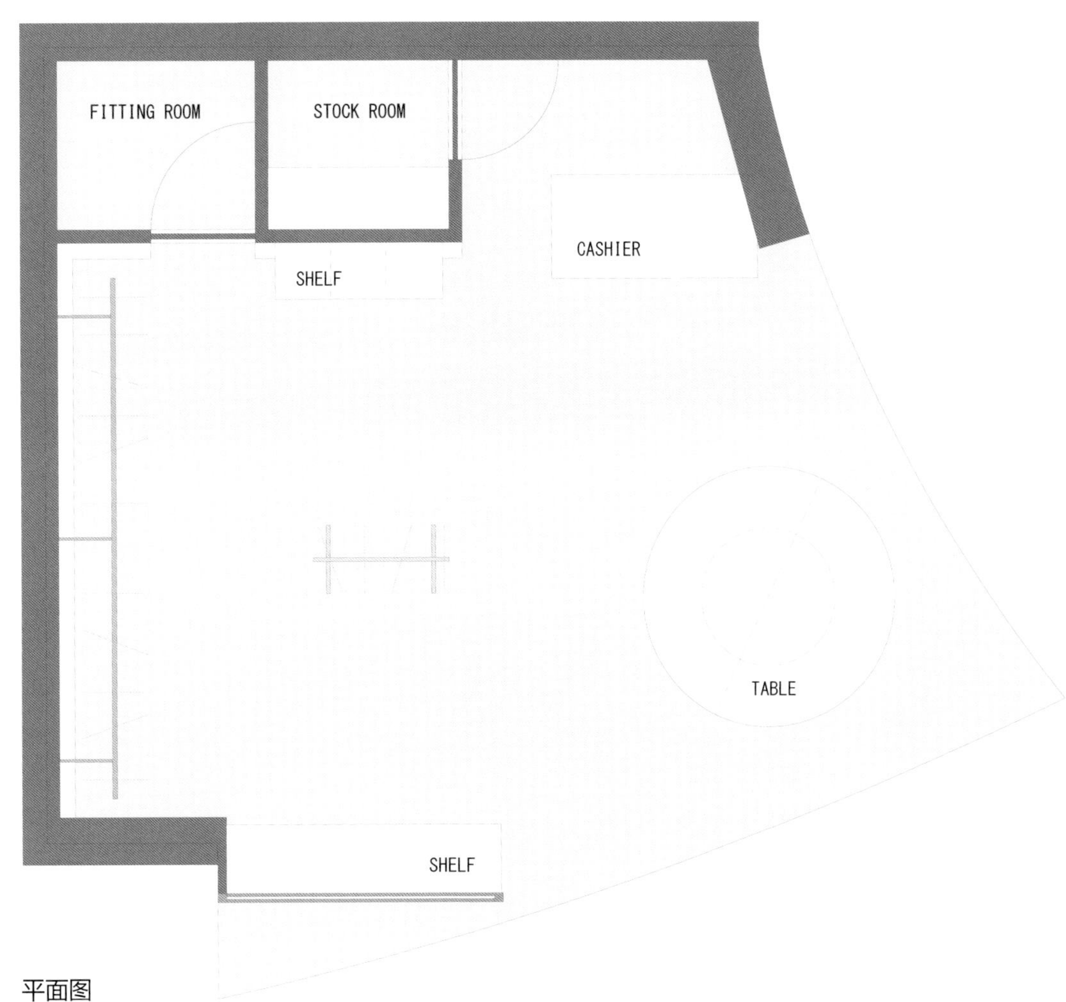

平面图

BAOBAO
ISSEY MIYAKE
PLEATS PLEASE

PLEATS
PLEASE
ISSEY MIYAKE
BAO BAO
ISSEY MIYAKE

PLEATS
PLEASE
ISSEY MIYAKE

Issey Miyake

三宅一生专卖店

Design agency: MOMENT	设计单位：Moment 设计事务所
Designer: Hisaaki Hirawata, Tomohiro Watabe	设 计 师：Hisaaki Hirawata、Tomohiro Watabe
Location: Japan	项目地点：日本
Area: 176 m^2	项目面积：176 m^2
Photography: Fumio Araki	摄　　影：Fumio Araki

This shop is blank space. It is quite important to bring out the charm of the products of ISSEY MIYAKE, which is famous in the world. I often think that there are too many decorations which are irrelevant to the products inside the store. What I strongly hope is that I want customers to see their collection in high quality inside chaste space. Therefore, I eliminate unnecessary decorations. What I focus on is how the clothes are displayed. That is to say, I attempt to design the store where customers are able to face the products in very simple atmosphere. For instance, there is a rich blank space on the facade as well.

As the height of the show window is reduced, the logo richly stays in a blank space. Also torsos in colorful fashion catch passersby's attention because of the contrast with minimal design on the facade. The long way from the entrance into the inside is like a runway, and it would excite customers as entering. The hanger rails hung from 3.8 meters mildly come into view during shopping. They make the hung clothes stand out inside minimum designed space.

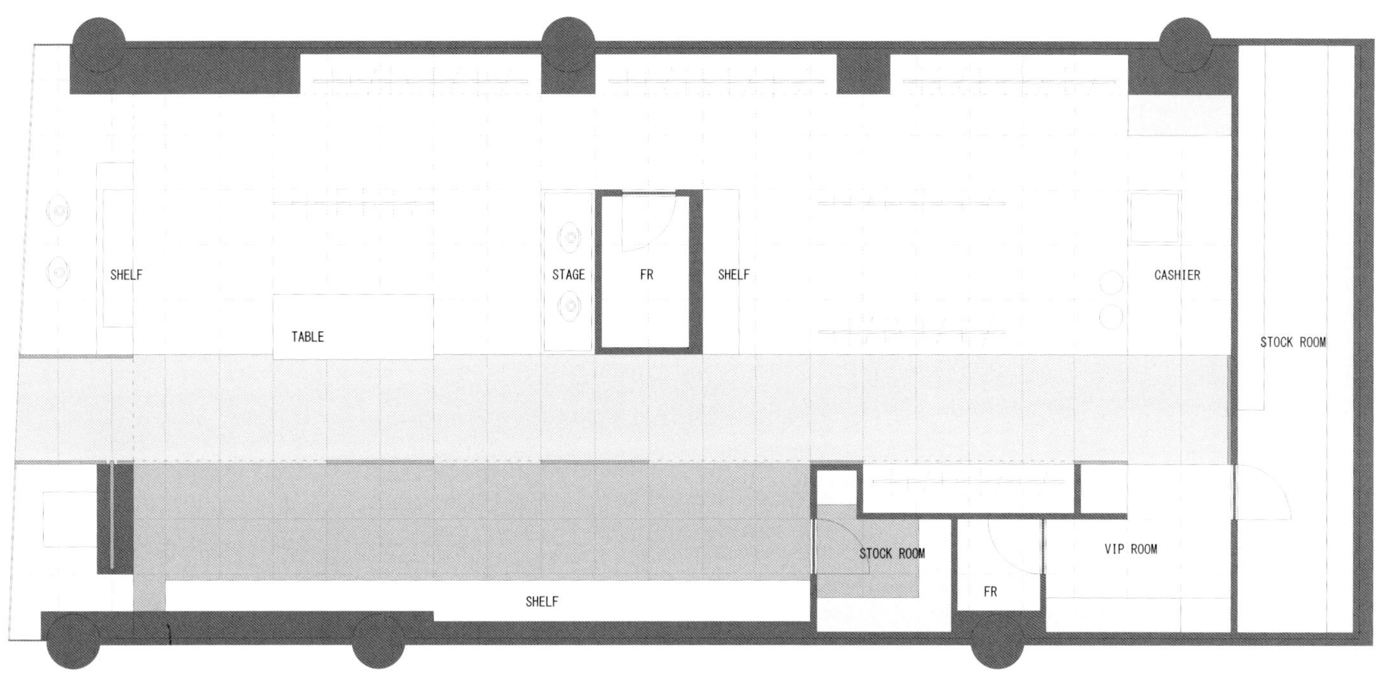

平面图

这是一个留白的空间，能将世界名牌三宅一生的产品魅力展现出来才是最关键。我经常都在想：其实很多店内的装饰品都与展示的产品没有丝毫关联。而我希望给顾客一个高质量的纯净空间来挑选喜爱的物品，所以在这个设计中，我剔除了所有不必要的装饰品。设计师所关注的是如何展现服饰，换句话说，能让顾客直接面对产品的简单氛围才是我所想要设计的店面空间，例如：一个内外都高度留白的空间。

由于橱窗的高度有所降低，品牌的商标摆放在了大气的门楣上。也正是因为外立面采取了最小化设计，才让橱窗内模特身上的鲜艳时装能够准确地抓住路人的视线。一条狭长的小路一直从门口延伸到店铺里面，如同一条跑道，引导顾客进入店内。进入店内，从 3.8m 高的顶棚处悬挂下来的挂衣架尤为显目，这些挂衣架的使用保证了室内设计的最小化。

Mynt Store

Mynt 旗舰店

Design agency: Dear design.
Art Director: Ignasi · Llauradó
Designer: Paulina Calcagno, Sebastián Pereyra
Location: Barcelona, Spain
Area: 40 m^2
Photography: Xavi Torrent

设计单位：Dear design 设计事务所
艺术总监：因格纳斯 · 劳拉多
设 计 师：Paulina Calcagno、Sebastián Pereyra
项目地点：西班牙巴塞罗那
项目面积：40 m^2
摄　　影：Xavi Torrent

The store design is based on a three-dimensional grid that creates a visually permeable volume, which commands the space, while generating niches to expose the product. The elements of the structure were progressively grown according to a recurrence relation inspired by the Fibbonacci sequence suggesting a progressive and open expansion of space.

The grid sets a dedicated exhibition place up for each accessory, although it grants the versatility of the showroom, the display elements can be rearranged with the arrival of new products. Lighting works as compositional element that emphasizes the rhythm of the grid and a mirror on the ceiling optically enlarges the space.

As we approach, the facade gradually discloses inside of the shop, creating a variety of different perspective views and diverse experiences in each spot. The shop window has a double reading, as the exhibition can be seen both from outside and inside, multiplying the effectiveness of exposure meters. A space at the back side of the store serves as storage space and staff facilities. Materials are used in there purest state: natural plaster slab, cement-like floor, wood. Wood and white glaze finishes create a set of contrasts exposing the texture of the original material.

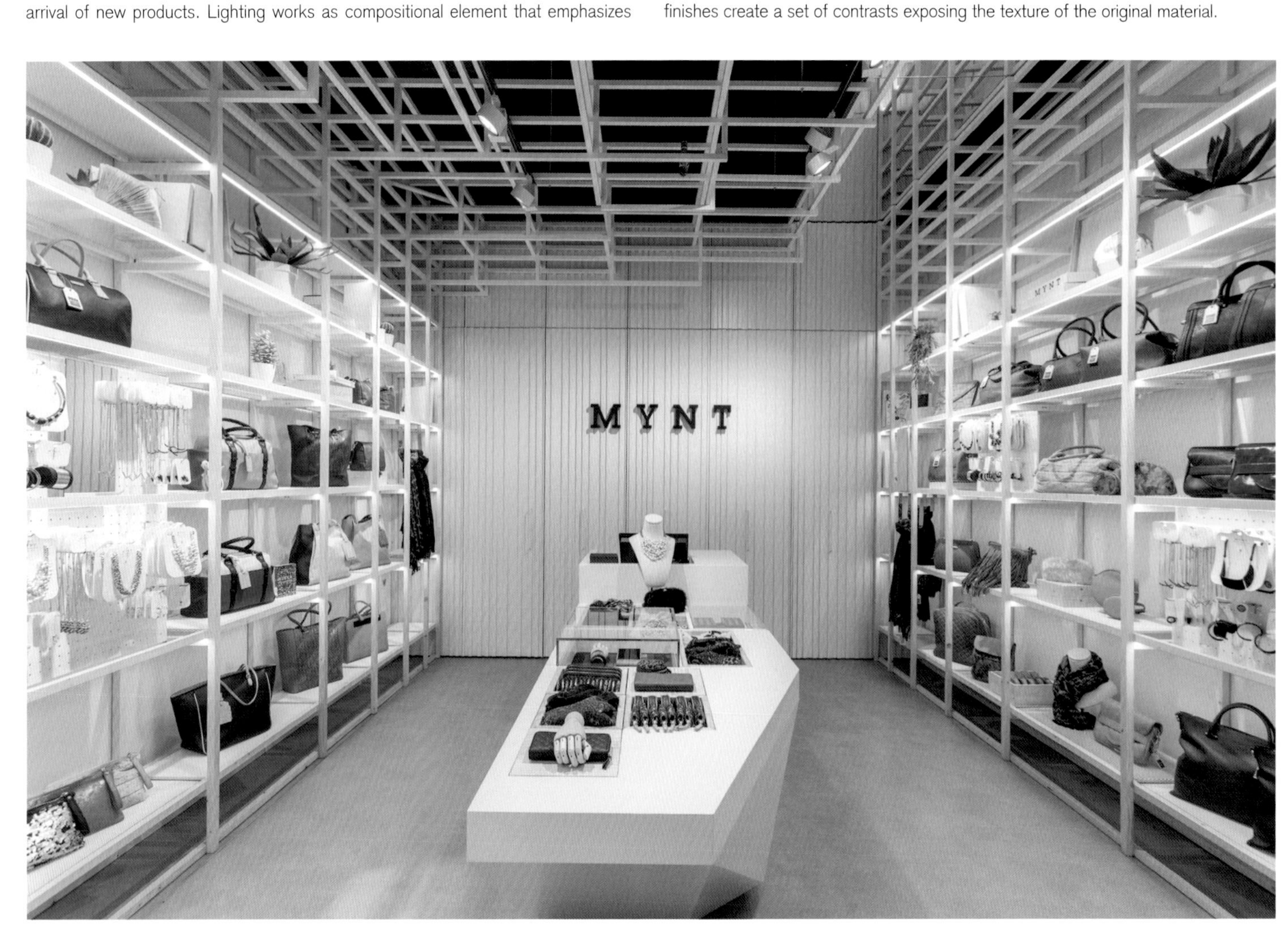

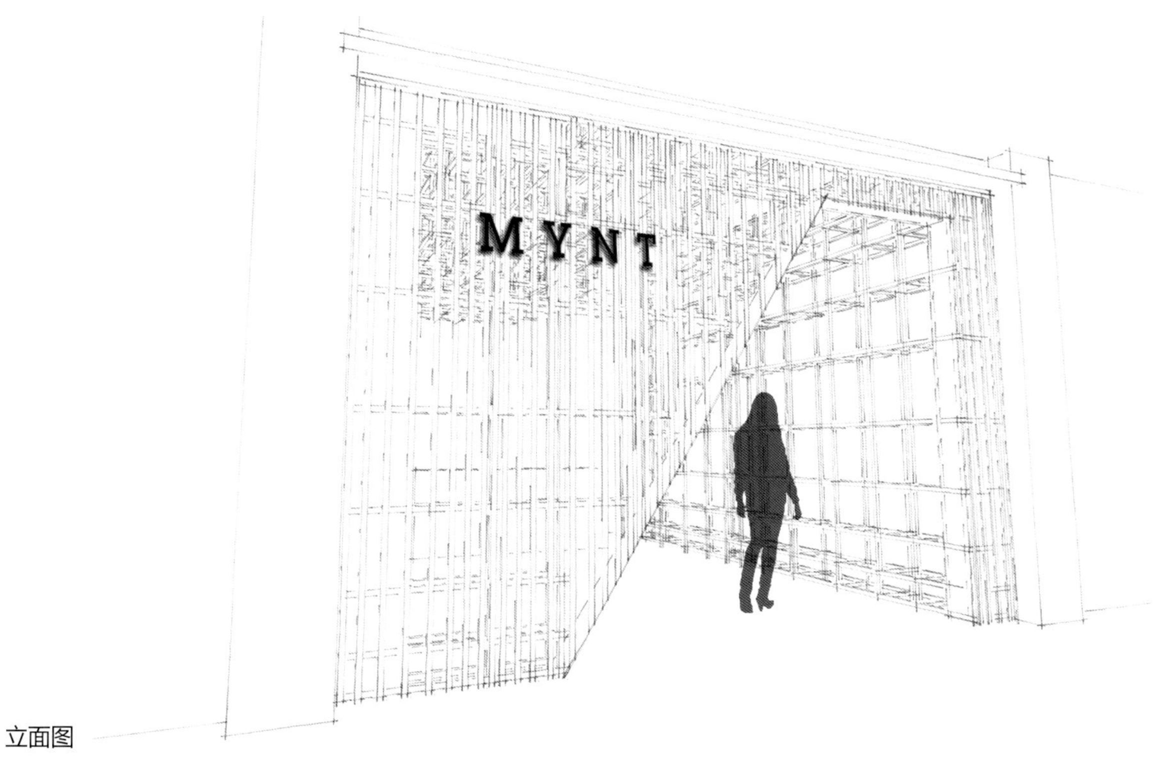

立面图

店内的设计是基于一个可视的三维柱体网格结构引导整个空间，以此来展示产品。这种结构元素是受到斐波纳契数列的启发，以逐步增长的方式营造一种开放的空间。

网格的设置使得每个物件都有自己特有的展览区，并且可用新款替换下旧款，赋予整个陈列室多样化特色。灯光加强了网格格调，顶棚上的镜子在视觉上扩大了整个空间。

慢慢靠近店铺，店内的全景逐渐展现在眼前，在不同的地点，可以有不同的多角度视图和多样化的视觉体验。从店内外均可以看到橱窗内展示的产品，增加了有效曝光率。店铺后面部分则是作为存储空间和员工活动区。内部装饰材料一共有三种：天然石膏板、水泥和木头。木头的原始质地和白色釉面形成鲜明对照。

Camper Together

Camper Together 品牌鞋店

服装鞋帽
Clothing shoes and hats

Design agency: Atelier Marko Brajovic	设计单位：AMB 设计事务所
Designer: Atelier Marko Brajovic	设 计 师：阿特利尔·马可·布拉加维克
Project collaboration: arch. Lelia Arruda Takeda	项目合伙人：Lelia Arruda Takeda
Location: Sao Paulo, Brazil	项目地点：巴西圣保罗
Photography: Fernando Laszlo	摄　　影：Fernando Laszlo

For our first project for Camper Together, we discovered our inspiration in traditional folkloric Brazilian festivities, where involving environments are created by dense layers of colored stripes. Using that strategy with shoe laces we modeled different shapes that implement in our space a sensitive and stenographic experience. Atelier Marko Brajovic Camper together is a model of collaboration between Camper and leading designers to create exclusive products and outstanding stores. Together responds to a new international reality that requires the capacity to integrate through design, different cultures and creative know-how into a single project together with and organization capable of communicating and distributing unique initiatives to a select global marketplace.

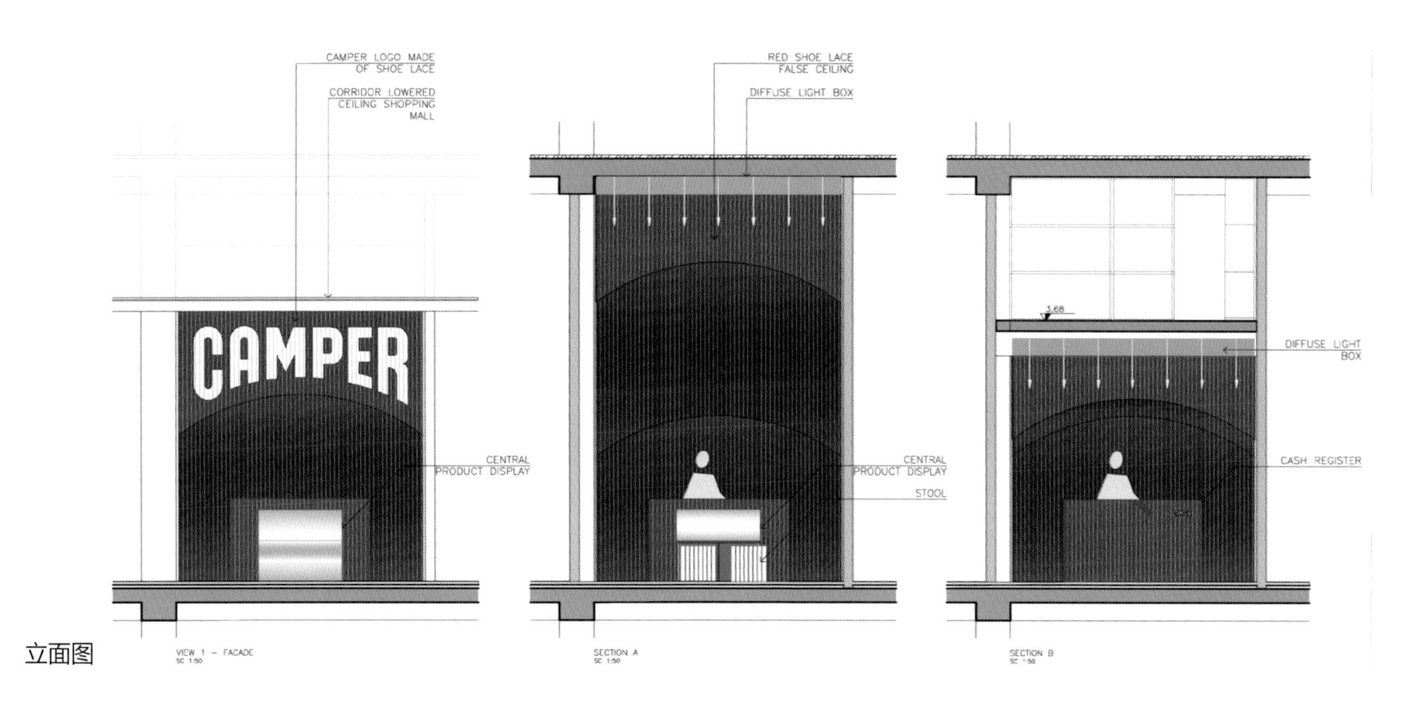

立面图

这是我们为 Camper Together 设计的第一个项目，其灵感来自巴西传统的节日民俗，使用有致密层的彩色条纹创建相关环境。我们模拟鞋带不同的形状在空间的敏感和效果图的体验。

Camper Together 是一个由露营者和设计师之间的合作模式创造的独特的产品和优秀的商店。在新的国际现实情势下，需要整合的能力，通过设计，将不同的文化和创意知识纳入同一个项目，针对特定的全球市场，具备独特的沟通与配送组织能力。

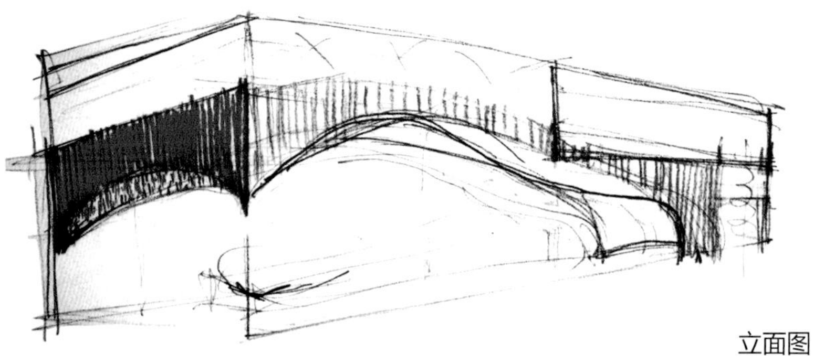

立面图

Skagen Store

斯卡恩时尚品牌店

Design agency: UXUS	设计单位：UXUS 设计事务所
Location: London	项目地点：英国伦敦
Area: 90 m^2	项目面积：90 m^2
Photography: Michael Franke	摄　　影：迈克尔·弗兰卡

UXUS developed a global retail platform for Skagen Denmark, part of Fossil Inc. The first two stores opened in Westfield London and Westfield Stratford City. The stores draw inspiration from Skagen's homeland, an eponymous town at the tip of Denmark, where land and sea meet in an eternal horizon line.

The displays for jeweler, timepieces and leather goods seem to form a landscape along the horizon, reminiscent of the ships and roofs dotting the coastline of Skagen. All of these individual pieces create a panorama for exploration, much like the many personalities that give the town of Skagen its unique character. Guided by Danish love for simplicity and "human touch" in craftsmanship, all fixed elements are made-up of clean lines and contrasted by soft, curved forms that add homely touches.

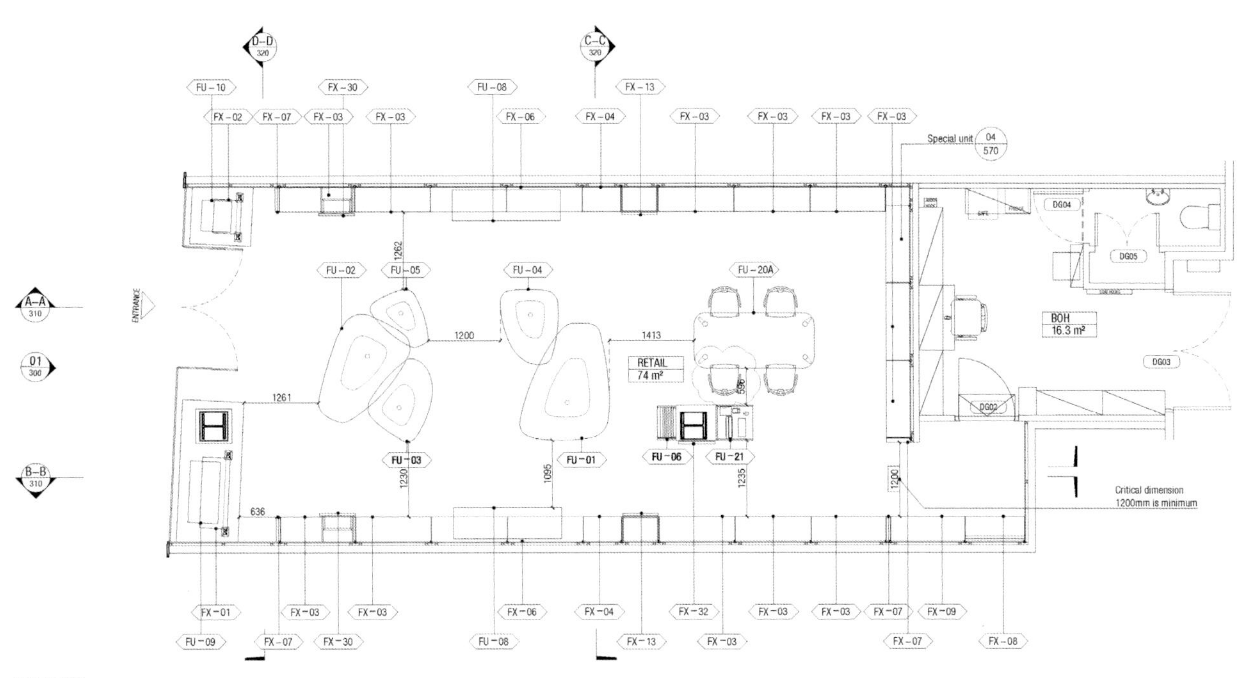

平面图

UXUS 为化石公司旗下子品牌 — 丹麦斯卡恩开创了一个全球零售平台，最初两家分店开在伦敦和斯特拉福德的韦斯特菲尔德商场里面。店铺设计的灵感来自于斯卡恩的故乡——一个位于丹麦顶端的同名小镇，在那个地方海天形成一线，地平线无限延伸。

店内珠宝、钟表和皮革商品的摆设看起来像沿着地平线形成的景观，让人不禁联想到轮船和点缀着斯卡恩海岸线的屋顶。所有这些单独的个体合在一起创建出一个全景，极具探索性，与那些塑造了斯卡恩小镇独特个性的特质有着异曲同工之妙。出于丹麦人对简单、有人情味的手工艺品的喜爱，店内所有的固定元素都采用简洁的线条，并由柔和的弯曲造型为空间增添几丝家的温馨。

Skagen
NANO HOUSE
DESIGNS

1000

Par La Roy Fashion Boutique

Par La Roy 时尚精品店

服装鞋帽
Clothing shoes and hats

Design agency: Savvy Studio	设计单位：Savvy 设计事务所
Design Collaboration: Emilio Álvarez	协同设计师：Emilio Álvarez
Location: Mexico	项目地点：墨西哥
Area: 80 m^2	项目面积：80 m^2
Photography: Alejandro Cartagena	摄　　影：亚利杭德罗・卡塔特纳

Par La Roy is a fashion boutique that promotes selected brands and upcoming artists and designers alike. The space is defined by a simple and clear layout which allows products to be displayed in an orderly and attractive fashion; its window displays showcase ever-changing trends, styles and products. The ambience itself is quite neutral, predominantly white, using only desaturated colors as complements, thus promoting a harmonious relation amongst the varied proposals that Par La Roy has to offer. The project's formal and structured language is based on two contrasting materials: metallic tubing and fabric. The tubes provide a set structure whilst the fabrics create volume, generating a unique and inviting atmosphere.

Par La Roy 是一个服装精品品牌，品牌创始人是一个艺术家和设计师，设计也以极高的要求来对待。该店被定义为一个简单明了的布局使产品陈列显眼且极具有吸引力，橱窗的展示风格和产品也是随着趋势的变化而不断变化，专卖店的环境设计是以中性、白色为主，局部使用不饱和的颜色作为辅助色。项目的正式结构化语言是基于两种互相对立的材料：金属管材和织物面料。金属管作为结构，面料形成体量，产生一个独特和诱人的气氛。

LAROY

PARLAROY

Harrolds Flagship

哈罗德旗舰店

Design agency: Landini Associates
Location: Melbourne, Australia
Area: 800 m^2
Photography: Trevor Mein

设计单位：Landini Associates
项目地点： 澳大利亚墨尔本
项目面积：800 m^2
摄　　影：特莱弗•米恩

Harrolds Menswear engaged Landini Associates to create a space that would showcase their luxury offering in top quality menswear and suits.
Using a warm palette, Landini Associates highlights luxury through details. The space needed to cater for a range of high end customers who would appreciate warmth, luxury and the finer things in life. The reasonable space allocation with magical effect of colors and lights creates a comfortable and warm atmosphere. While the carpet and magnificent furniture adds nobility and elegance to the whole space.
"Guys, you have helped us create the best menswear store in Australia, our customers are telling us that too." – Theo Poulakis, Harrolds Menswear .

受哈罗德男装的邀请，Landini Associates 为其旗舰店打造了一个用以展示高端男装的低调奢华的空间。该案以暖色调为主，特别强调从细节处展现奢华的感觉。该店的服务对象均为高端人士，生活中较为注重高雅品位、奢华品质以及温暖感。各个功能区域的合理划分，以及对色调和灯光的妙用，营造出舒适温暖的感觉。铺了地毯的地面，辅以古典华丽的家具，为整个空间增添了几分高贵典雅。
业主 Theo Poulakis 表示对这个设计特别满意。他说：“这一定是澳大利亚最棒的男装店面！就连我们的顾客都这么说。”

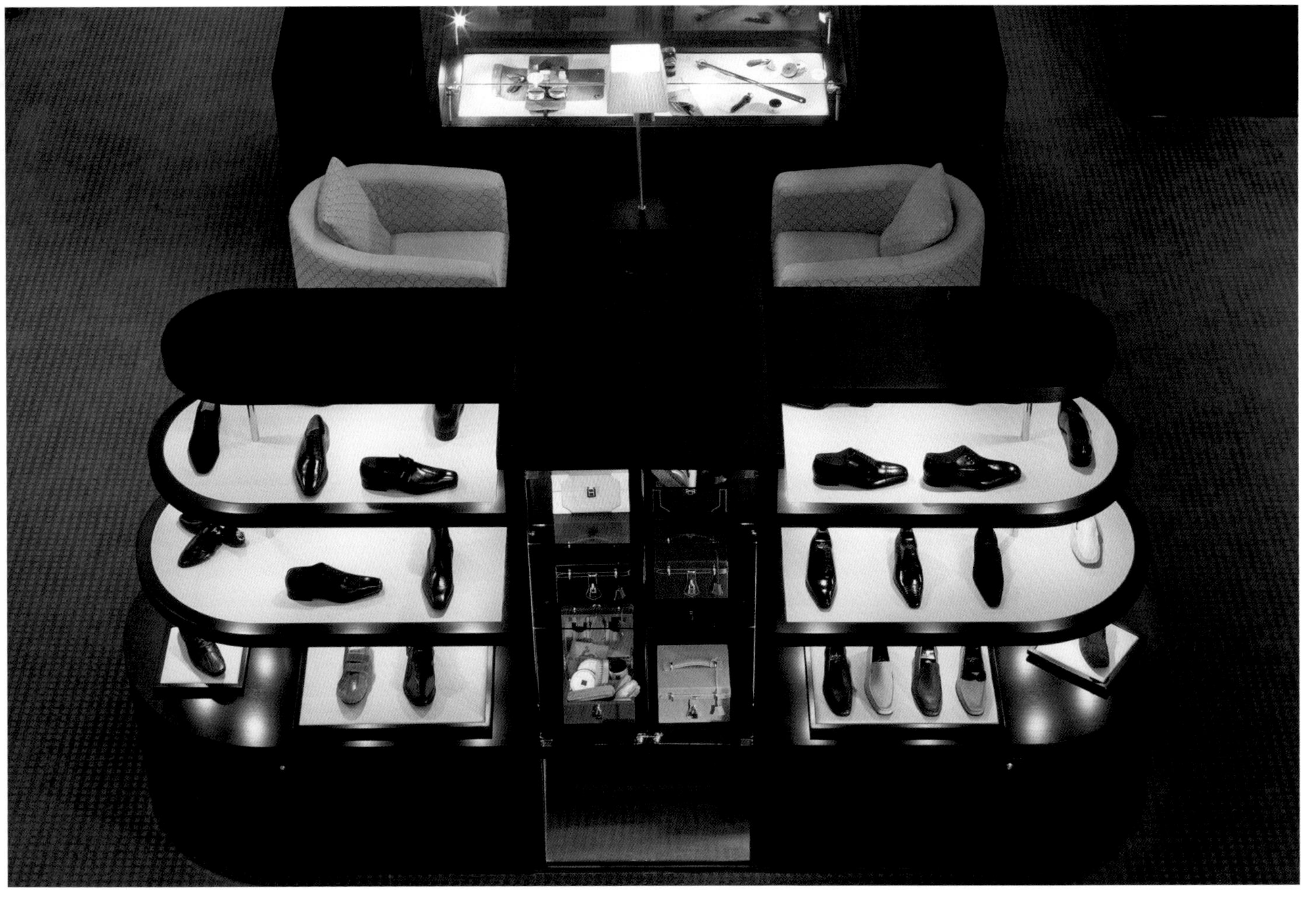

Zhuang Zi Store

庄姿服饰

Design agency: Xiamen YMLT Design Decoration Engineering Co., Ltd.
Designer: Zeng Weikun, Zeng Weifeng, Li Lin
Location: Xiamen, Fujian
Area: 97 m^2
Photography: Liu Tengfei

设计单位：厦门一亩梁田设计顾问
设 计 师：曾伟坤、曾伟锋、李 霖
项目地点：福建省厦门市
项目面积：97 m^2
摄　　影：刘腾飞

The store design flexibly applies checker elements in many parts of the space including the wall and the ceiling, adding more sense of depth and landscape visions to the whole space. The main color of the design is pure white, along with soothing yellow, and supplemented with calm black, which creates a comfortable and magnificent atmosphere, lacking in no elegance. As for the hanging racks, the square steel tubes echo with other checker elements, realizing the comparison between black and white. The function of ornament is infinite and intensive which succeed in gaining mutual transformation between richness and simplicity.

本案为品牌服装店的店面设计，“方格”元素灵活运用于墙面和顶棚等多个部分，使整个空间增添了不少的跳跃性与景观视觉。空间内的色调是运用纯净的白色、搭配温和的黄色、零星点缀沉静的黑色营造出一个大气舒适的空间，却不缺乏稳重与典雅。衣架的装饰，用细而修长的方形钢管与“方格”元素形成呼应，黑与白的对比，无限且密集的装饰功能，实现丰富与单纯之间的转化与统一。

More&less Furniture Shanghai M50 store

多少家具上海 M50 本店

家居 household

Design agency: Shanghai Mooma design agency / Shanghai Shanxiang Architecture Design Co. Ltd	设计单位：上海善祥建筑设计有限公司 上海木码设计机构
Designer: Wang Shanxiang, Gong ShuangYan	设 计 师：王善祥、龚双艳
Location: Shanghai	项目地点：上海
Area: 208 m^2	项目面积：208 m^2
Photography: Hu Wenjie	摄　　影：胡文杰

More& Less furniture is a brand established by the famous furniture designer Mr Hou Zhengguang, which mainly designs, produces and sells original furniture and household items with contemporary Chinese cultural features, the majority of products take solid wood as materials. This is its flagship shop. The store locates in Shanghai's most famous creative industry park. It is on the first floor, in a deep small alley, which is not easy to be found. The main area is on the first floor with an interlayer inside as storage and product packing room. The store was once a workshop, the internal part retains the original concrete frame of a plant.

More& Less furniture has a slogan: it is cozy to be hidden in the house, which matches the sentence is a small house graphics with the strong sense of painting. Thus, two "homes" with small house shape are embedded in the rectangular plane of shop to constitute the main interest of the space. The first "home" is arranged at the entrance, which is very eye-catching and can be seen in alley entrance at first glance. The second "home" is laid in the most internal shop. One is in front and the other one is at back, one is large and the other is small. Small houses are painted white, namely no color, in order to foil furniture. Meanwhile, white is outstanding in the park dominated by the gray tone. The height of internal exhibition space is 4 meters. The hay color straw wallpaper is attached to most of the walls, which soften coldness and hardness of the concrete.

多少
MORELESS

多少家具是由著名家具设计师侯正光先生创办的品牌，主要设计、生产和销售具有当代中国文人色彩的原创家具及家居用品，以实木居多，这是其旗舰店。店面坐落于一层，在一个较深的小弄堂里，并不好找。主要面积在一层，内部有一个小夹层，作为储藏及产品包装间。店铺原为厂房，内部保留了当初厂房的混凝土框架。

多少家具有一句广告语：小隐于宅，而且这句话配有一个很具有绘画感的小房子图形。于是，在长方形店铺平面里嵌入两个小房子形的“家”，构成了空间的主要趣味。第一个“家”布置在入口，在弄堂口一眼便可望见，十分醒目。第二个“家”布置在店铺的最内部。一前一后，一大一小。小房子粉刷成白色，也即没了颜色，以烘托家具。同时白色在灰色调为主的园区内也比较出挑。展厅内部空间高度为4m，大部分墙面贴了干草色的草编墙纸，柔化了混凝土的冷硬。

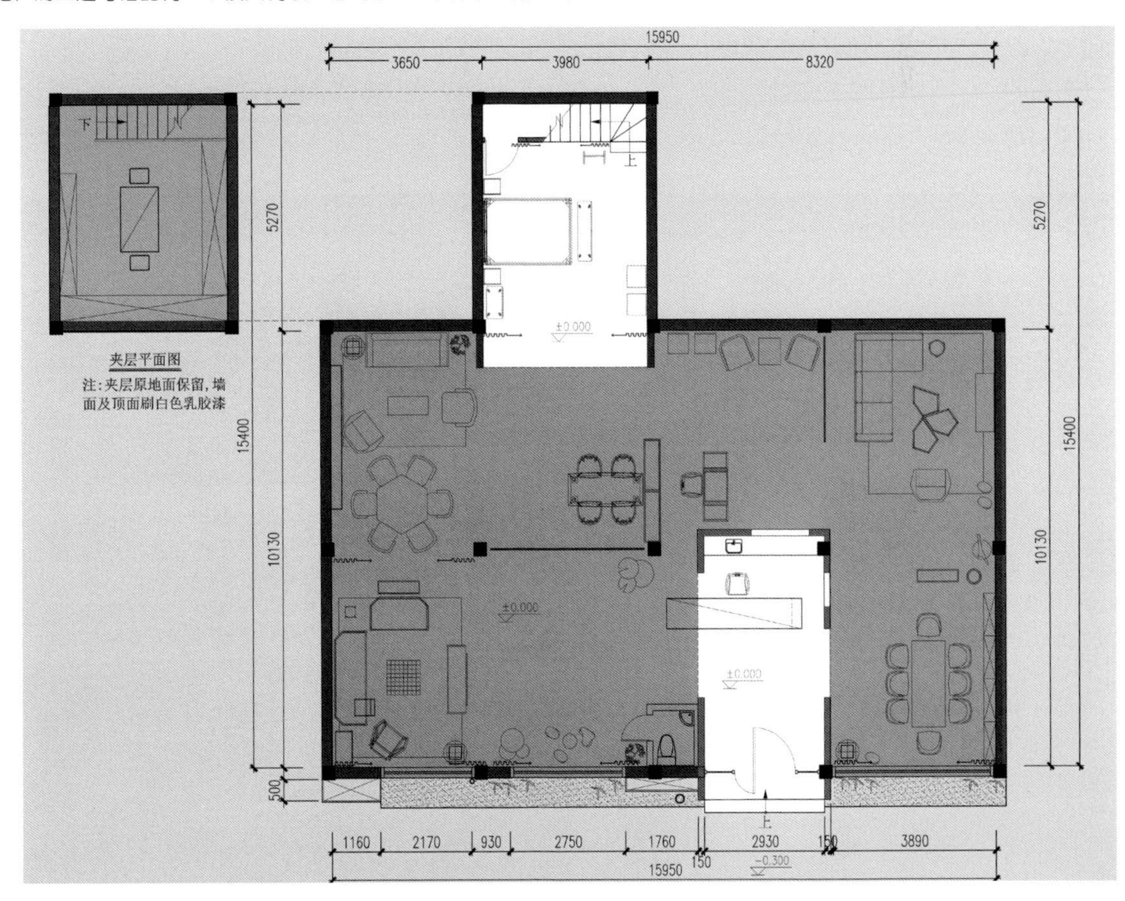

平面图

Ming Gu Yuan Gold Phoebe Furniture Chamber

茗古园——金丝楠木汇馆

家居
household

Design agency: Hongkong Dayu Space Design Co., Ltd.
Designer: Chen Jie
Location: Fuzhou, Fujian
Photography: Zhou Yuedong

设计单位：香港大于空间设计有限公司
设 计 师：陈　杰
项目地点：福建省福州市
摄　　影：周跃东

Located in a shabby and small alley, the Ming Gu Yuan Gold Phoebe Furniture Chamber is particularly tranquil. When the night falls, the cloud carved on the upside of the façade is gleaming in the darkness. The exterior façade made of transparent glass makes the interior space an impressive visible card. In the shape of a circle hole, the door connects the exterior to the interior, which is traditional and classic. The hallway wall is decorated with various Chinese building components, enriching the space level. On the right side of the aisle is the main space of the chamber, still using the original ornaments, with a structure at a downward angle to draw the visitors' attention to the gold phoebe furniture. This area is designed to display the furniture made of fresh gold phoebe and behind it there is another area displaying the study furniture made of old gold phoebe. The room next to the area has collections on display except for gold phoebe furniture, and most of them have stories with their owners.

茗古园——金丝楠木汇馆位于汉唐文化城对面的小巷子里，在这个略显衰败的街道里，店面显得格外清幽典雅。夜幕时分，店面上方的浮云图案泛着光泽，散发着一种若即若离的气息。汇馆的外立面用透明的玻璃材质，让内部空间的景致成为一张鲜明的视觉名片。

汇馆的门以园洞的形式存在，内外的视觉衔接让空间拥有了十足的亲和力，并潜移默化地将传统意味铺成开来。进门后的门厅墙以各式中式建筑构件作为装饰，不同的纹理质感和层层叠叠的组合丰富了墙面的层次。走道的右侧是汇馆的主题空间，上方的装饰延续着老建筑楼梯构件的装饰，向下的结构倾向使得人们的视觉焦点自然而然地落到区域中的金丝楠木家具上。这个空间展示以金丝楠木新料为主制作的家具。这个空间的背后设置了一个书房家具的展示间，以金丝楠老料为主。与之毗邻的房间，除了金丝楠木家具外，还陈列了其他属性的收藏品，这些新奇的收藏，都与主人有着这样或那样的缘分。

Corleone Solon

柯里昂美发店

Design agency: 7980 Interior Design Studio
Designer: Yu Xiaoqi
Location: Beijing
Area: 200 m^2

设计单位：北京七九八零室内设计工作室
设 计 师：于晓祺
项目地点：北京
项目面积：200 m^2

The case is located in Sanlitun SOHO in Beijing, whose owner has just come back in china from European for studying. He really loves foreign decorative styles and details, and wishes to implant these into this case. Thus, designers take requirements of the owner into consideration, abandoning the complicated texture and decorations in European designs, but simplified lines according to the situation and understandings on classic European style.

Kermesinus wooden furnishing is the first choice for this case, while the yellow floor along showing a sense of European classic style, which makes this space full of nobility and elegance. Moreover, luxury crystal lamps are hanging in the centre of the space to enhance the gorgeous atmosphere. Leather sofa echoes with the concise lines in the space, showing a spacious European space. As a result, plenty of returnees like the owner come into the store frequently.

本项目位于北京三里屯 SOHO，店主欧洲留学回国，对国外的装饰风格与细节极有情结，并且希望将这种风格移植到中国。设计师体会修正了店主的要求，摒弃了欧式设计过于复杂的肌理和装饰，简化了线条并根据自己的经验以及对欧式古典的理解加以改造。

本案在设计上，选用了以暗红色为主的实木家具，以米黄色的地板搭配，烘托了欧式复古的情怀，使整个空间具有端庄典雅的贵族气质。另外又添加了欧式的吊灯，将奢华宫廷的情怀发挥到极致。皮质的沙发与整个空间简洁的线条交相呼应，凸显出欧式的大气。随着装修后的经营，很多和店主相似经历的“海归”成为了这家店铺的长期拥趸。

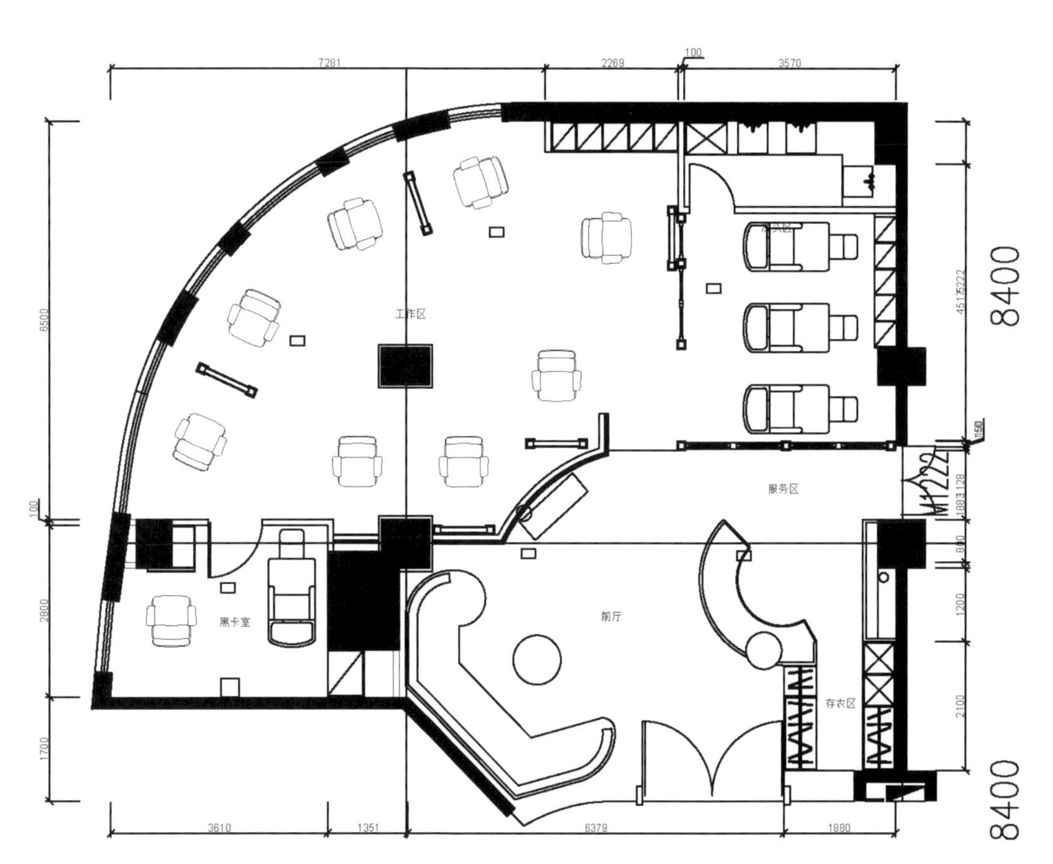

平面图

Olin Salon

欧灵造型

Design agency: Xiamen YMLT Design Decoration Engineering Co., Ltd.
Designer: Zeng Weikun, Zeng Weifeng, Li Lin
Location: Xiamen, Fujian
Area: 320 m^2
Photography: Liu Tengfei

设计单位：厦门一亩梁田设计顾问
设 计 师：曾伟坤、曾伟锋、李 霖
项目地点：福建省厦门市
项目面积：320 m^2
摄　　影：刘腾飞

The Olin Salon is a Chinese style hair salon. Breaking the convention of the narrowly defined Chinese style, the designers refine the concept, instead of following the retro completely. Modern materials are used, such as aluminium, glass and granite. To keep traditional visual images, the walls are covered with grey cement fiberboards, along with the façade decorated with the latticed screen, creating a clean and simple space of Modern Chinese style. The main color of the interior space is grey, showing a low-key but noble space temperament. The reasonable spatial layout with good partitions enables the visual space to extend, which is elegant and magnificent.

本案为中式主题发型沙龙空间设计。跳脱出狭义的"中国风"，设计师没有完全遵循复古的路线，而是经过一番提炼简化，采用现代材料，如金属铝材、玻璃、花岗石等。室内设计通过墙面的灰色水泥纤维板以及立面的装饰花格屏风来保留传统视觉意象，创造出一个现代新中式的干净利落的空间和氛围。室内色调以沉稳的灰色为主，彰显低调而不失高贵的空间气质。空间布局合理，隔而不断，令人的视线得以延伸，大气中透露着典雅。

Aquira Hair Salon

Aquira 美发沙龙

Design agency: Love the Life
Designer: Akemi Katsuno; Takashi Yagi
Location: Kanazawa-ku, Kanagawa, Japan
Area: 58.5 m^2
Photography: Shinichi Sato

设计单位：Love the Life 工作室
设 计 师：勝野明美、八木孝司
项目地点：日本神奈川金泽区
项目面积：58.5 m^2
摄　　影：佐藤新一

This hair salon was on the second floor of a small building. This area is a residential suburb of Yokohama, and there is a railroad station in the neighborhood.

We imagined a story that the client creates wonderful hairstyles with the help of angels. The hut-shaped wooden frames colored with bright orange were placed in the center of floor, and the pendant lights of special order were attached to each hut. The lights symbolized angels. Six styling chairs were installed. Half of them inherited from the former barber.

The outside of the building was painted white, and store's name logo was placed on top. The wooden frame of the same as the interior was placed around the logo. Orange lines and round shaped fluorescent lamps constitute impressive symmetric appearance.

该美发店位于一栋小型建筑的二楼，此处是横滨的一个住宅区，临近火车站。根据客户的需求，设计师提出了用天使这一设想来创造美丽的发型设计。小屋形的木质框架涂以鲜亮的橙色，被放置在空间的中央，吊灯也被按照独特的秩序悬挂在木屋上面。灯光象征着天使。室内摆放了六张独具品位的椅子，其中一半是由前理发师遗留下来的。

建筑的外立面被漆成白色，上面还写上了美发店的名称，在标志的周边围绕着与内部小屋形一样材质的木质结构。外立面上橙色线条和圆形荧光灯对称搭配，无疑是令人印象深刻的组合。

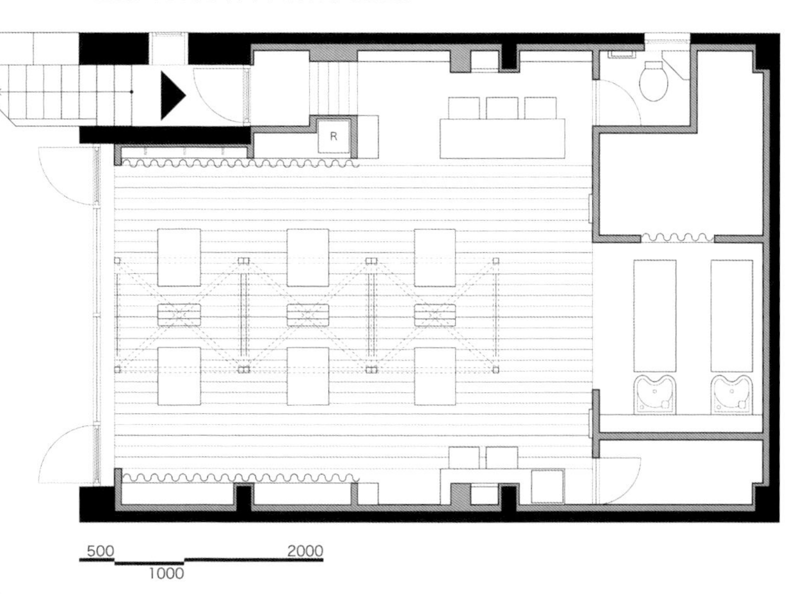

平面图

Fit Hair Salon

Fit 美发沙龙

Design agency: Love the Life
Designer: Akemi Katsuno & Takashi Yagi
Location: Yokohama-shi, Kanagawa, Japan
Area: 56 m^2
Photography: Shinichi Sato

设计单位：Love the Life 工作室
设 计 师：勝野明美、八木孝司
项目地点：日本神奈川金泽区
项目面积：56 m^2
摄　　影：佐藤新一

The owner is managing a hair design studio at the center of Yokohama City, and "Fit" is the second branch. He wanted to offer good service that adjusted to local populace's needs and a state-of-the-art hair design here. "Fit" is at the ground floor of a small apartment in a residential quarter near the central part of the city.

We wanted to reproduce the memory of waterside in this place. Mirrors were used as metaphor of surface of the water. The guest has the hair set while looking into the surface of the water. The size and the position of the mirror were strictly decided based on the experiment by the original dimension. Mirror units of three sets (six seats in total) were arranged along the blue belt painted on the center of the floor. The space of the remainder was designed extremely simply. Wild grasses were planted at the terrace.

客户在横滨市中区经营着一家美发工作室，Fit 是其第二家店，他想要在这里塑造一个适应当地人民需求的艺术型美发沙龙。该沙龙位于城市中心一个住宅区小公寓内的首层。

设计师试图重现水乡的记忆，采用镜子隐喻水面，让顾客能够在做头发的时候看着水面。镜子的尺寸和位置都严格按照原有尺寸来设计。每 3 块镜子组成一组镜面单元，共有 6 个座位，都沿着地板上画出的蓝色带状区域分布。空间内剩余的空间设计都非常简单，阳台上栽种了不少野草。

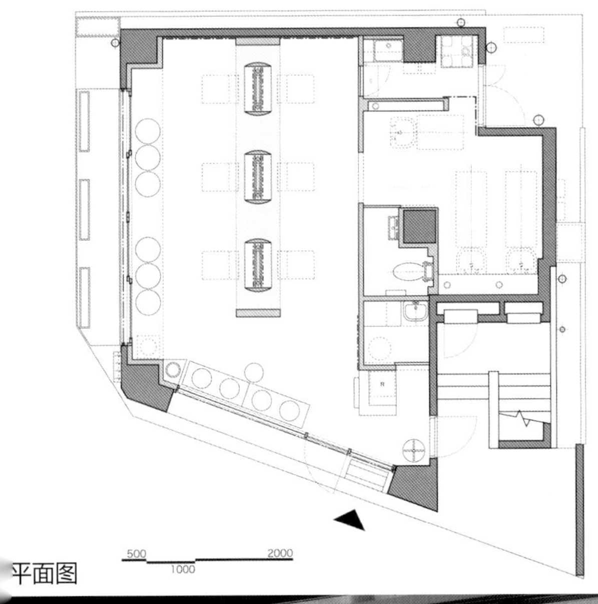

平面图

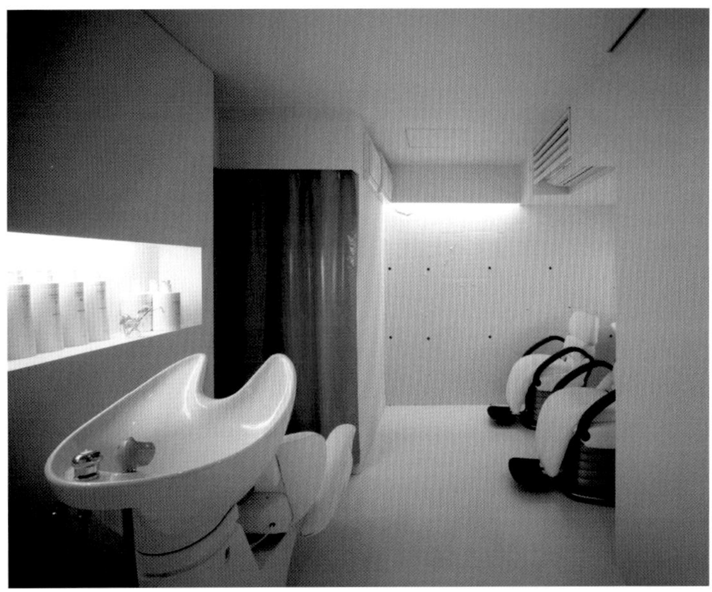

Premiera, Bangkok, Thailand

曼谷 Premiera 珠宝专卖店

Design agency: Design Worldwide Partnership 设计单位：DWP 设计公司
Location: Bangkok, Thailand 项目地点：泰国曼谷
Area: 200 m^2 项目面积：200 m^2

World-class architecture and interior design firm DWP were commissioned to create the interiors and store façade for Premiera, a luxury jewelry store, within the high-end shopping mall Siam Paragon in Bangkok's city centre.

DWP was responsible for conceiving and setting the scene for this new branded store, developed as a younger and hip brand, differentiating from its mother company Premier Diamond. DWP custom designed the displays and jewelry fixturing components, feature walls, transaction area, lounge area, grand VIP room and back of house and office areas. In addition, DWP was tasked with the shop front design, in coordination with Siam Paragon and their guidelines, plus the security systems, including but not limited to CCTV, strong room and entrance. The overall effect is a chic, unique and contemporary sophistication, with warmth and elegance, where silver, bronze and dark finishes allow the diamonds to shine as the stars. Specialist lighting was employed, to ensure optimum vibrancy, as well as high gloss surfaces, natural silks in display cases.

PREMIERA
FIRE EXIT

EXQUISITE JEWELLERY

世界级顶尖建筑及室内设计公司DWP接到委托，为一家奢侈珠宝店一Premiera设计店铺室内装修及门店设计，该店位于曼谷中心城区的暹罗典范高端购物广场。

DWP将设计一个全新的品牌店，从而塑造与Premier钻石母公司不一样的年轻潮流品牌的形象。店内的展示工具及和珠宝固定元件都是特别定制的，同时还设置了独具特色的墙面、交易区、休息区、豪华贵宾室以及办公室等。

除了店面设计之外，DWP还负责与配合暹罗典范广场，按照规定加上安全系统，包括但不限于闭路电视、保险库及出入口的安全。整体效果展现出的不仅仅是一种时尚、独特、现代化的复杂性，也是温暖优雅的，而钻石更是在银饰品、黑色家具的衬托下如星星般闪耀迷人。室内采用专业的照明，以确保最佳视觉效果，突出展示柜中的天然丝绸以及光滑表面。

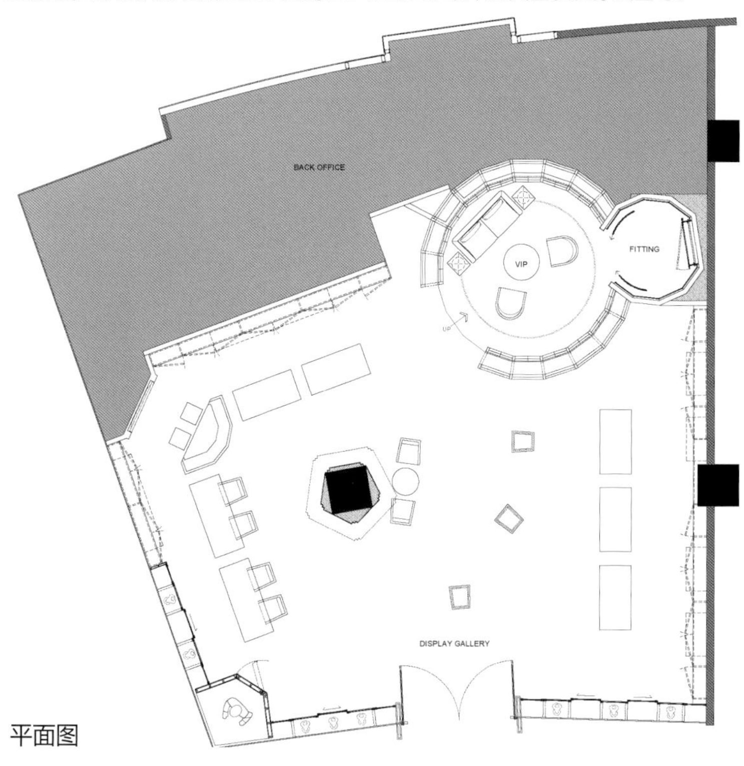

平面图

Link Jewelry Store Concept

Link 珠宝概念店

珠宝 首饰
Jewelry

Design agency: MINAS KOSMIDIS (ARCHITECTURE IN CONCEPT)	设计单位：MINAS KOSMIDIS 设计事务所
Location: Thessaloniki, Greece	项目地点：希腊塞萨洛尼基
Area: 30 m²	项目面积：30 m²
Photography: Studiovd N.Vavdinoudis - Ch.Dimitriou	摄　　影：Studiovd N.Vavdinoudis - Ch.Dimitriou

Link means hoop, connect, unite. The deconstruction of the word link gave the designers a big amount of line segments, vertically, horizontally and pairs of segments which are combined at an angle. The rapid and continuous repetition of the word and the effort of writing the word link with clasped letters, like a stroke of the pen, a constant zigzag was the element of inspiration in designing the space. This is how a 30 sqm place was designed, the main element- symbol is the thin crooked line which is sometimes the base support of furniture and sometimes electricity metal links.

As entering the store from the right hand side we can see the main furniture where the jewelry are displayed and stored which is the parallelepiped section of the left side of the store that is missing from the wooden paneled walnut which is after the floor. The sharp marble - level table (serviced), rounded by a golden metal blade like a precious stone, who wants to become jewelry come up. In the opposite side and the same logic of the row stone there are two mirrors posted on the top and left part of the furniture-table display of the jewelry. All the other surfaces were covered with aggregates in neutral color. The whole scene comes alive with the presence of the painting - photograph by a known photographer.

Link 的概念是环绕、相连和混合。通过进一步解析 link 这个词，设计师们受到了一些启发，在设计上采用了很多线条状的装饰，这些线条或水平或垂直，或者以某个角度相连起来。整个设计给人的感觉就像是有人大笔一挥，在空中重复写下拆分的单词 link，并以 Z 字形式排列。设计师别出心裁地用这些曲折的线段装饰这个只有 30 平方米的空间，它们或者立在地板上当做家具的支撑，或者以金属装饰的形式悬在头顶装点空间。

从右侧进入商店，一眼就能看到展示珠宝的主展柜设置在店内左侧，这是一个平行六面体结构的空间，给人的视觉感受仿佛是其末端融入了胡桃木的地板，渐渐消失了一样。这部分空间看似是由 Z 字钢架支撑，实则这些钢架都是旁边桌子的桌脚，而这个金属包边的大理石桌如同一块珍贵的宝石一般，静待采撷。另一边的墙上挂了两面镜子，下边的展柜陈列着各式珠宝，而墙上挂的那幅出自名家之手的摄影作品，在墙面中性色调的烘托下，使整个空间都变得鲜活灵动起来。

LINK

Media Arts Library Cafe

图书馆式休闲吧—日本文化厅媒体艺术祭

其他
The others

Design agency: TORAFU ARCHITECTS
Location: Tokyo, Japan
Area: 194 m^2
Photography: Takumi Ota

设计单位：TORAFU 建筑设计事务所
项目地点：日本东京
项目面积：194 m^2
摄　　影：太田拓实

We built an event space intended to promote media arts which opened during the 16th Japan Media Arts Festival. We envisioned a space bringing three functions together: a library presenting prized mangas from the festival's Manga Division along with related works, a panel exhibition introducing initiatives directed at promoting media arts and an event space featuring talks with creators and researchers from every field.

The venue is made of volumes resembling big building blocks that can be used as bookshelves, panels or benches, etc., to match these three different functions. We gave shapes the versatility to be able to combine the panel and bench volumes and use bookshelves as benches during events. Moreover, we gave color to the space by posing needle punched carpets on the seating surface of benches and the sides of bookshelves.

We strived to make the entire space catchy to the visitors' eyes looking through the glass at the impressive three-dimensional stacked volumes.

本案是专门为第 16 届日本文化厅媒体艺术祭设计的活动场地。设计师欲将场地打造成三合一的空间，包括展示漫画部提供的珍贵漫画和相关作品的展示区，用于介绍多媒体艺术项目的展示区，以及一个供创作者和各领域专家们交谈的活动区。不同的区域放置了不同的大积木块，它们被用作书架、展板和长凳。设计师运用他们的巧思，赋予这些积木块可通用的功能，如在活动期间，展板和书架都可用作休息的凳子。另外，他们甚至在这些被当做凳子、书架的积木块上铺上一层颜色各异的地毯，既装点了整个空间，又让室内变得明亮而多彩。外墙面采用透明玻璃的设计，让进门的客人一眼就能看到室内的景象，从而被吸引，给人留下深刻的印象。

オーラルヒストリー　杉井ギサブロー監督
4
闇の国々
岳 みんなの山
ましろのおと

6
7
平成24年度
第16回 文化庁メディア芸術祭
マンガ部門
受賞作品
新人賞 凍りの掌 シベリア抑留記
新人賞 千年万年りんごの子
新人賞 ぼくらのフンカ祭

Primera

芙莉美娜化妆品专卖店

其他
The others

Design agency: Landini Associates	设计单位：Landini Associates
Location: Korea	项目地点：韩国
Area: 300 m^2	项目面积：300 m^2
Photography: Courtesy of AmorePacific	摄　　影：Courtesy of AmorePacific

Landini Associates collaborated with Korean super brand Amore Pacific to launch their eco-friendly skin care and cosmetic brand Primera into the retail market by developing an international brand strategy, new identity, packaging and retail design concept that are now being rolled out across Asia.

Natural materials are used to reinforce the natural ethos at the core of Primera's brand philosophy. Key features include a beauty counseling table, sprouting energy tester bar and a special services area where customers are invited to enjoy tea and a massage. As always, Landini's integrated approach to design has created a holistic customer experience for the Korean market.

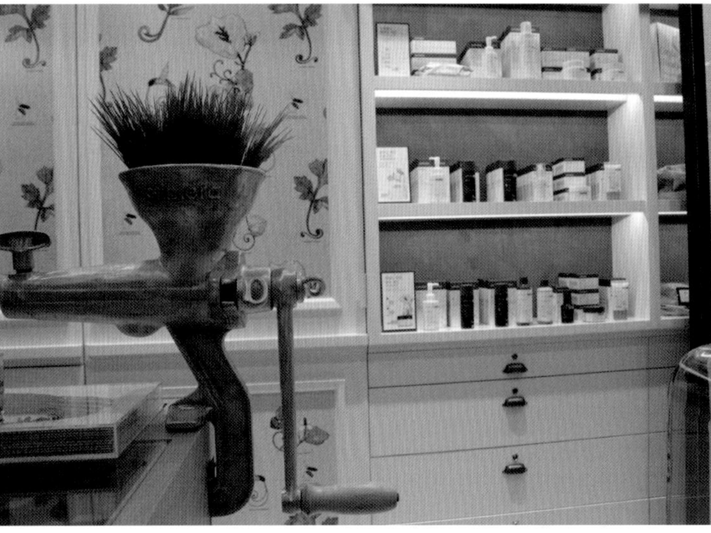

Landini Associates 与韩国第一化妆品牌爱茉莉太平洋公司合作，致力于通过改进国际品牌策略、身份认同、包装和零售设计概念等将自然生态的化妆品牌芙莉美娜投入零售市场。自然气质是芙莉美娜的品牌宗旨，因此室内的设计围绕这一点，采用了各种自然材料。室内设了不同分区，有美容咨询台、能量测试吧，还设了一个特别服务区，在这里客人可以边享受茶点边咨询。室内空间的色调以纯净的白色和柔和的淡黄色为主，一如该品牌化妆品一般温和自然，素雅又不失明亮温暖。印了碎花的墙纸和木质的家具为整个空间增添了几分古典优雅的感觉。整体的设计为前来的每一位客户都带来完整舒适的体验。

Fish-man Tea Shop

淘海人茶货铺

Design agency: Xiamen YMLT Design Decoration Engineering Co., Ltd.	设计单位：厦门一亩梁田设计顾问
Designer: Zeng Weikun, Zeng Weifeng, Li Lin	设 计 师：曾伟坤、曾伟锋、李　霖
Location: Xiamen, Fujian	项目地点：福建省厦门市
Area: 33 m^2	项目面积：33 m^2
Photography: Liu Tengfei	摄　　影：刘腾飞

The Fish-man Tea Shop is located in the Zengcuoan cultural and educational tourist area that used to be a fishing village. Inspired by the fishing boat, the designer uses the boat decks as the decorative finishes of exterior walls whose mottled and antique texture feels as natural as the original one. The shop is clean and tidy. The designed tea canisters are divided and placed separately by different colors. Extending to the exterior space, the floor is covered with emery and cement. Like tea leaves of tea bushes planted here, the whole space is pristine and natural.

淘海人茶货铺

福建茶 南洋情 淘海梦

本案位于厦门曾厝垵文教旅游区，这里原本是个临海的小村庄，故店名为“淘海人”。淘海即出海捕鱼之意，设计师以捕鱼的渔船为设计灵感，将渔船的船板作为店面的外墙饰面，其斑驳的纹理与仿古质感如“土生土长”般融合于街道边，自然得像是原本就在这里。室内整洁干净无需赘言。茶叶罐经过设计后以色块区分排列，整齐而又纯粹。地面用金刚砂水泥作为室外的延伸。整个空间如同茶叶一般淳朴自然。

AER Telecom Store

爱施德 AER 通信概念零售店

其他
The others

Design agency: COORDINATION ASIA	设计单位：协调亚洲建筑设计工作室
Designer: Tilman Thuermer	设 计 师：逖尔曼 · 图蒙
Location: Shenzhen	项目地点：深圳
Area: 150 m^2	项目面积：150 m^2
Photography: Tilman Thuermer	摄 影：逖尔曼 · 图蒙

COORDINATION ASIA completed a new breed of telecom stores named AER for AISIDI, one of China's leading resellers for mobile and digital products and services. AER is a retail brand that enhances the life of the individual mobile user by offering customized mobile services in a playful, cool and customer-focused environment. COORDINATION ASIA took on branding as well as VI and store design for AER.

Inspired by the growing importance of mobile Internet use in daily life in China, COORDINATION ASIA created an out-of-the-box brand that turns the act of purchasing a mobile product into an active and fun experience. AER is based on a great understanding of mobile lifestyle, in which mobile devices keep you connected, entertained and updated through a variety of online and offline APPs.

The store is designed as an interactive environment that caters to the needs of different target customers: Trendy, Lifestyle and Tech Savvy. Products are thematically presented in combination with related accessories, APPs and carriers on custom-made presentation tables with 'serving trays'. Following a black runway from the entrance, customers find the APP Bar where they can try out mobile APPs on a large interactive screen. Painted pegboards are used to cover the walls of the store. This allows display shelves to be hung freely within the whole space and to be changed location at all times. The whole interior is flexible, so that changes in layout can be made quick and easily.

COORDINATION ASIA has created a concept that fits perfectly the needs of the mobile-minded customer, who is used-to "click and play" and never wants to be bored. The ever-changing store concept of AER will keep customers coming back, providing them with an exciting and new experience every time.

随着移动互联网络在中国人的日常生活中变得日益重要，协调亚洲受此启发为 AER 通信概念零售店营造了一个突破陈旧框架束缚的氛围，将移动产品的购买行为转化成一种互动式的有趣体验。AER 的创立完全基于对移动生活潮流的深入理解——移动设备通过丰富多样的线上线下应用软件将你与周围联系起来，提供娱乐，并不断更新。互动的环境设计重点面向三类不同客户的需求：最流行、最经典、最前沿。一条黑色走道从店面入口处通向尽头的一个大型互动屏幕，应用迷你吧是其核心，客户可在此现场试用移动应用程序。店面布局从整体上灵活展示数字云的混合多变形式，可在整个空间内任意悬挂展示架和装饰品，并可快速灵活地改变悬挂装饰的位置。协调亚洲的动态店面设计理念首次完美呈现了以移动生活为重心，习惯于“点击，开始游戏”，且不喜欢沉闷无趣的顾客的需求，而 AER 不断变化的店铺概念将提升客户回头率，使他们每次进店均能获得精彩纷呈的全新体验。

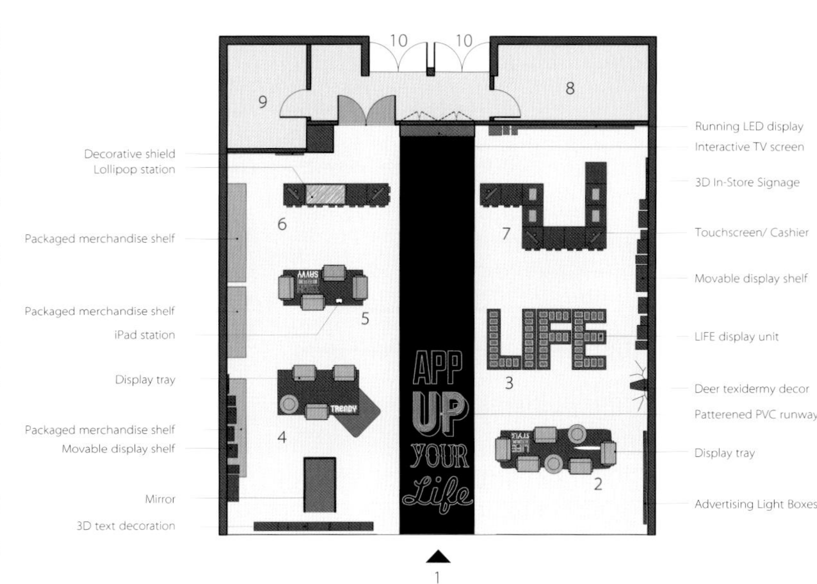

平面图

TAP
APP
UP
YOUR
Life

TAP
深圳 享易无限 酷动数码 爱施德

DRESS
IT
UP
装扮靓丽
APP
UP
YOUR

Yan Ji You Book Store
北京言几又书店

其他
The others

Design agency: Kyle Chan & Associates Design
Designer: Kyle Chan
Location: Beijing
Area: 754.5 m^2
Photography: Wong Kin Fai

设计单位：峻佳室内设计
设 计 师：陈峻佳
项目地点：北京
项目面积：754.5 m^2
摄　　影：黄建辉

"Book is everywhere" is not only the brand tenet but also the design concept of Yan Ji You Book Store, based on which, the designer applies elements of book to the whole design. Natural materials such as wood, brick and concrete are used to create a comfortable and natural atmosphere. The color of furniture made of wood adds simplicity and freshness to the space, along with the black and yellow color.

A notable characteristic is the spiral stair whose handrails consist of book shelves. It seems that the stair is made of various books. The design changes the original space structure, by making the stair in the front place to highlight the store theme and using the stair to connect the two floors and enable the interior to interact with the exterior space. The wall covered with books is a big attraction, on which the light is projected in the pattern of the store name Yan Ji You to create a concave-convex feeling. Echoing with the aroma of coffee, the book store becomes an elegant cafe.

The concrete platform and stuffs made of pristine materials highlight the simplicity and elegance of the store as well as its unique feature. The horn-shaped droplight hanging from the ceiling shows the good style and taste.

"书．无处不在"是言几又其中一个品牌宗旨，同样也是这个项目的设计概念。基于此，我们希望以书作为设计的主角，把书的元素融入空间内。同时利用木、砖、水泥等自然物料营造出舒适、自然的感觉。室内的家具摆设及特色装置以木为主要物料，配合黄色及黑色，增加简约及年轻的感觉。把各个元素融合为一，令整体空间更具生活品位及性格。

设计的一大特色便是颇具时代感的旋转楼梯。楼梯以书籍作为扶手的一部分，看起来就像一座以书砌成的楼梯；更改原有的店面结构，把旋转楼梯设于店面的前方位置，突出主题的同时让其变成书店的一大亮点；并且以楼梯贯通两层的空间，加强室内室外空间的互动性。另一个特色便是咖啡厅内的书墙，把书籍安装于墙上并于书上投射品牌口号，刻意强调书墙的凹凸感，使书卷气息更浓厚，配合咖啡厅的咖啡香气，大大提升生活品位。而书店方面，为了突出商品的特色，以水泥地台等简洁的物料及手法衬托出空间的简约感。只在顶棚吊挂一个喇叭造型吊灯，令空间带有一点点艺术品位，把格调提升。

整个空间采用了不少自然物料、运用现代、简约的手法给平平无奇的书店带来艺术品位及型格感觉。

立面图

café
Yan Ji You
Grande
Tall
Venti
Venti
Grande

图书在版编目（CIP）数据

商业店铺设计［购物篇］/深圳市海阅通文化传播有限公司主编．— 北京：中国建筑工业出版社，2015.2
ISBN 978-7-112-17751-6

Ⅰ.①商··· Ⅱ.①深··· Ⅲ.①商业－服务建筑－室内装饰设计－世界－图集 Ⅳ.①TU247-64

中国版本图书馆CIP数据核字(2015)第027091号

责任编辑：费海玲　张幼平　王雁宾
责任校对：李美娜　赵　颖
装帧设计：龙萍萍
采　　编：刘太春

商业店铺设计［购物篇］
深圳市海阅通文化传播有限公司　主编
*
中国建筑工业出版社出版、发行（北京西郊百万庄）
各地新华书店、建筑书店经销
深圳市海阅通文化传播有限公司制版
北京方嘉彩色印刷有限责任公司印刷
*
开本：880×1230毫米　1/16　印张：6¾　字数：216千字
2015年9月第一版　2015年9月第一次印刷
定价：48.00元
ISBN 978-7-112-17751-6
(27023)